For instructor-assigned homework, **MasteringPhysics™** provides the **only adaptive-learning online tutorial and assessment system**. Based on years of research of how students work physics problems, the system is able to coach you with feedback specific to your needs, and hints when you get stuck. The result is targeted tutorial help that optimizes your study time and maximizes your learning.

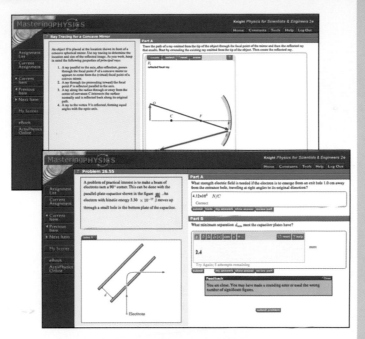

Mastering PHYSICS

If your professor requires **MasteringPhysics™** as a component of your course, your purchase of a new copy of the Pearson Addison-Wesley textbook already includes a Student Access Kit. You will need this Student Access Kit to register.

If you did not purchase a new textbook and your professor requires you to enroll in **MasteringPhysics™**, you may purchase online access with a major credit card. Go to **www.masteringphysics.com** and follow the links to purchasing online.

For self study, **MasteringPhysics™** provides you with **ActivPhysics OnLine™**, the most comprehensive library of applets and applet-based tutorials available. **ActivPhysics OnLine™** utilizes visualization, simulation, and multiple representations to help you better understand key physical processes, experiment quantitatively, and develop your critical-thinking skills. This library of online interactive simulations is coupled with thought-provoking questions and activities to guide your understanding of physics.

Activ
ONLINE
Physics

System Requirements:
(Subject to change. See website for up-to-date requirements.)
Windows: 250 MHz CPU; Microsoft® Windows 2000, ME, XP
Macintosh: 233 MHz CPU; Apple® Mac OS® 10.2, 10.3, 10.4

Both:
- RAM: 64 MB
- Screen resolution: 1024 x 768
- Browser (OS dependent. Check website for more detail):
  Firefox 1.5, 2.0; Internet Explorer 6.0, 7.0; Safari 1.3, 2.0

Browser settings and players/plug-ins: **Javascript** must be enabled in your browser. Any **popup blocker** should either be disabled or set to allow popups from: session.masteringphysics.com
Flash™ Player 9.0 or higher (downloadable from www.adobe.com).

For self study, **MasteringPhysics™** also provides you with a complete electronic version of your textbook. You can access the **eBook** from within MasteringPhysics™ and review or reference your online textbook anywhere, anytime. No additional software or viewer required.

Table of Problem-Solving Strategies

Note for users of the five-volume edition:
Volume 1 (pp. 1–477) includes chapters 1–15.
Volume 2 (pp. 478–599) includes chapters 16–19.
Volume 3 (pp. 600–785) includes chapters 20–25.
Volume 4 (pp. 786–1183) includes chapters 26–37.
Volume 5 (pp. 1140–1365) includes chapters 37–43.

Chapters 38–43 are not in the Standard Edition.

ActivPhysics OnLine™ Activities

ActivONLINE Physics   www.masteringphysics.com

VOLUME 2

PHYSICS

FOR SCIENTISTS AND ENGINEERS SECOND EDITION

A STRATEGIC APPROACH

RANDALL D. KNIGHT

CALIFORNIA POLYTECHNIC STATE UNIVERSITY, SAN LUIS OBISPO

PEARSON

Addison
Wesley

San Francisco Boston New York
Cape Town Hong Kong London Madrid
Mexico City Montreal Munich Paris
Singapore Sydney Tokyo Toronto

Publisher:	Adam Black, Ph.D.
Development Manager:	Michael Gillespie
Development Editor:	Alice Houston, Ph.D.
Project Editor:	Martha Steele
Assistant Editor:	Grace Joo
Media Producer:	Deb Greco
Sr. Administrative Assistant:	Cathy Glenn
Director of Marketing:	Christy Lawrence
Executive Marketing Manager:	Scott Dustan
Sr. Market Development Manager:	Josh Frost
Market Development Associate:	Jessica Lyons
Managing Editor:	Corinne Benson
Sr. Production Supervisor:	Nancy Tabor
Production Service:	WestWords PMG
Illustrations:	Precision Graphics
Text Design:	Hespenheide Design
Cover Design:	Yvo Riezebos Design
Manufacturing Manager:	Evelyn Beaton
Manufacturing Buyers:	Carol Melville, Ginny Michaud
Photo Research:	Cypress Integrated Systems
Director, Image Resource Center:	Melinda Patelli
Manager, Rights and Permissions:	Zina Arabia
Image Permission Coordinator:	Michelina Viscusi
Cover Printer:	Phoenix Color Corporation
Text Printer and Binder:	Courier/Kendallville
Cover Image:	Composite illustration by Yvo Riezebos Design; photo of spring by Bill Frymire/Masterfile
Photo Credits:	See page C-1

Library of Congress Cataloging-in-Publication Data
Knight, Randall Dewey.
   Physics for scientists and engineers : a strategic approach / Randall D. Knight.--2nd ed.
     p. cm.
   ISBN-13: 978-0-8053-2736-6
   1. Physics--Textbooks. I. Title.
   QC23.2.K654 2007
   530--dc22

                                             2007026996

ISBN-13: 978-0-321-51672-5
ISBN-10: 0-321-51672-9

PEARSON

Addison
Wesley

www.aw-bc.com

2 3 4 5 6 7 8 9 10—CRK—10 09 08 07

Brief Contents

v

About the Author

Randy Knight has taught introductory physics for over 25 years at Ohio State University and California Polytechnic University, where he is currently Professor of Physics. Professor Knight received a bachelor's degree in physics from Washington University in St. Louis and a Ph.D. in physics from the University of California, Berkeley. He was a post-doctoral fellow at the Harvard-Smithsonian Center for Astrophysics before joining the faculty at Ohio State University. It was at Ohio State that he began to learn about the research in physics education that, many years later, led to this book.

Professor Knight's research interests are in the field of lasers and spectroscopy, and he has published over 25 research papers. He also directs the environmental studies program at Cal Poly, where, in addition to introductory physics, he teaches classes on energy, oceanography, and environmental issues. When he's not in the classroom or in front of a computer, you can find Randy hiking, sea kayaking, playing the piano, or spending time with his wife Sally and their seven cats.

Preface to the Instructor

In 2003 we published *Physics for Scientists and Engineers: A Strategic Approach.* This was the first comprehensive introductory textbook built from the ground up on research into how students can more effectively learn physics. The development and testing that led to this book had been partially funded by the National Science Foundation. This first edition quickly became the most widely adopted new physics textbook in more than 30 years, meeting widespread critical acclaim from professors and students. In this second edition, we build on the research-proven instructional techniques introduced in the first edition and the extensive feedback from thousands of users to take student learning even further.

Objectives

My primary goals in writing *Physics for Scientists and Engineers: A Strategic Approach* have been:

- To produce a textbook that is more focused and coherent, less encyclopedic.
- To move key results from physics education research into the classroom in a way that allows instructors to use a range of teaching styles.
- To provide a balance of quantitative reasoning and conceptual understanding, with special attention to concepts known to cause student difficulties.
- To develop students' problem-solving skills in a systematic manner.
- To support an active-learning environment.

These goals and the rationale behind them are discussed at length in my small paperback book, *Five Easy Lessons: Strategies for Successful Physics Teaching* (Addison-Wesley, 2002). Please request a copy from your local Addison-Wesley sales representative if it is of interest to you (ISBN 0-8053-8702-1).

FIVE EASY LESSONS
Strategies for Successful
Physics Teaching

RANDALL D. KNIGHT

Textbook Organization

The 43-chapter extended edition (ISBN 0-321-51333-9/978-0-321-51333-5) of *Physics for Scientists and Engineers* is intended for a three-semester course. Most of the 37-chapter standard edition (ISBN 0-321-51661-3/978-0-321-51661-9), ending with relativity, can be covered in two semesters, although the judicious omission of a few chapters will avoid rushing through the material and give students more time to develop their knowledge and skills.

There's a growing sentiment that quantum physics is quickly becoming the province of engineers, not just scientists, and that even a two-semester course should include a reasonable introduction to quantum ideas. The *Instructor Guide* outlines a couple of routes through the book that allow most of the quantum physics chapters to be included in a two-semester course. I've written the book with the hope that an increasing number of instructors will choose one of these routes.

- **Extended edition,** with modern physics (ISBN 0-321-51333-9/978-0-321-51333-5): Chapters 1–43.
- **Standard edition** (ISBN 0-321-51661-3/978-0-321-51661-9): Chapters 1–37.
- **Volume 1** (ISBN 0-321-51662-1/978-0-321-51662-6) covers mechanics: Chapters 1–15.
- **Volume 2** (ISBN 0-321-51663-X/978-0-321-51663-3) covers thermodynamics: Chapters 16–19.
- **Volume 3** (ISBN 0-321-51664-8/978-0-321-51664-0) covers waves and optics: Chapters 20–25.
- **Volume 4** (ISBN 0-321-51665-6/978-0-321-51665-7) covers electricity and magnetism, plus relativity: Chapters 26–37.
- **Volume 5** (ISBN 0-321-51666-4/978-0-321-51666-4) covers relativity and quantum physics: Chapters 37–43.
- **Volumes 1–5** boxed set (ISBN 0-321-51637-0/978-0-321-51637-4).

The full textbook is divided into seven parts: Part I: *Newton's Laws*, Part II: *Conservation Laws*, Part III: *Applications of Newtonian Mechanics*, Part IV: *Thermodynamics*, Part V: *Waves and Optics*, Part VI: *Electricity and Magnetism*, and Part VII: *Relativity and Quantum Mechanics*. Although I recommend covering the parts in this order (see below), doing so is by no means essential. Each topic is self-contained, and Parts III–VI can be rearranged to suit an instructor's needs. To facilitate a reordering of topics, the full text is available in the five individual volumes listed in the margin.

Organization Rationale: Thermodynamics is placed before waves because it is a continuation of ideas from mechanics. The key idea in thermodynamics is energy, and moving from mechanics into thermodynamics allows the uninterrupted development of this important idea. Further, waves introduce students to functions of two variables, and the mathematics of waves is more akin to electricity and magnetism than to mechanics. Thus moving from waves to fields to quantum physics provides a gradual transition of ideas and skills.

The purpose of placing optics with waves is to provide a coherent presentation of wave physics, one of the two pillars of classical physics. Optics as it is presented in introductory physics makes no use of the properties of electromagnetic fields. There's little reason other than historical tradition to delay optics until after E&M. The documented difficulties that students have with optics are difficulties with waves, not difficulties with electricity and magnetism. However, the optics chapters are easily deferred until the end of Part VI for instructors who prefer that ordering of topics.

What's New in the Second Edition

This second edition reaffirms the goals and objectives of the first edition. At the same time, the extensive feedback we've received from scores of instructors has led to numerous changes and improvements to the text, the figures, and the end-of-chapter problems. These include:

- More streamlined presentations. We have shortened each chapter by one page, on average, by tightening the language and reducing superfluous material.
- Conceptual questions. By popular request, the end of each chapter now includes a section of conceptual questions similar to those in the *Student Workbook*.
- Pencil sketches. Each chapter contains several hand-drawn sketches in key worked examples to provide students with explicit examples of the types of drawings they should make in their own problem solving.
- New and revised end-of-chapter problems. Problems have been revised to incorporate the unprecedented use of data and feedback from more than 100,000 students working these problems in MasteringPhysics™. More than 20% of the end-of-chapter problems are new or significantly revised, including an increased number of problems requiring calculus.

Significant chapter and content changes include the following:

- Two-dimensional kinematics has been brought forward to Chapter 4, immediately following the chapter on vectors. This chapter also covers circular-motion kinematics in detail (rather than delaying circular-motion kinematics to the chapter on rotational dynamics) to give a more integrated understanding of kinematics.
- Newton's third law (Chapter 7) now immediately follows and is more closely linked to the chapter on dynamics in one dimension. Revised interaction diagrams are simpler to draw and conceptually more powerful.
- The mechanisms of heat transfer (conduction, convection, and radiation) have been included in Chapter 17 (Work, Heat, and the First Law of Thermodynamics).

- Spherical mirrors are now covered in Chapter 23 (Ray Optics), and the entirely new Chapter 24 (Optical Instruments) treats cameras, microscopes, telescopes, and vision. This is the only new chapter in the second edition.
- Dielectrics have been added to the section on capacitors in Chapter 30 (Potential and Field), and electric current (Chapter 31) now follows the presentation of electric potential.
- Some topics in Chapters 34 (Electromagnetic Induction) and 35 (Electromagnetic Fields and Waves) have been rearranged for a more logical presentation of ideas.
- Blackbody radiation and Wien's law have been added to Chapter 38 (The End of Classical Physics).

Pedagogical Features

Your Instructor's Professional Copy contains a 10-page illustrated overview of the pedagogical features in this second edition. The *Preface to the Student* demonstrates how these features are designed to help your students.

The Student Workbook

A key component of *Physics for Scientists and Engineers: A Strategic Approach* is the accompanying *Student Workbook*. The workbook bridges the gap between textbook and homework problems by providing students the opportunity to learn and practice skills prior to using those skills in quantitative end-of-chapter problems, much as a musician practices technique separately from performance pieces. The workbook exercises, which are keyed to each section of the textbook, focus on developing specific skills, ranging from identifying forces and drawing free-body diagrams to interpreting wave functions.

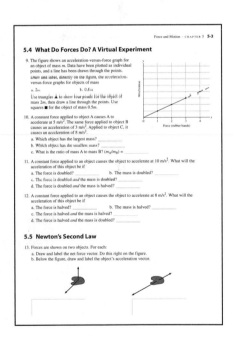

The workbook exercises, which are generally qualitative and/or graphical, draw heavily upon the physics education research literature. The exercises deal with issues known to cause student difficulties and employ techniques that have proven to be effective at overcoming those difficulties. The workbook exercises can be used in class as part of an active-learning teaching strategy, in recitation sections, or as assigned homework. More information about effective use of the *Student Workbook* can be found in the *Instructor Guide*.

Available versions: Extended (ISBN 0-321-51357-6/978-0-321-51357-1), Standard (ISBN 0-321-51642-7/978-0-321-51642-8), Volume 1 (ISBN 0-321-51626-5/978-0-321-51626-8), Volume 2 (ISBN 0-321-51627-3/978-0-321-51627-5), Volume 3 (ISBN 0-321-51628-1/978-0-321-51628-2), Volume 4 (ISBN 0-321-51629-X/978-0-321-51629-9), and Volume 5 (ISBN 0-321-51630-3/978-0-321-51630-5).

Instructor Supplements

- The **Instructor Guide for Physics for Scientists and Engineers** (ISBN 0-321-51636-2/978-0-321-51636-7) offers detailed comments and suggested teaching ideas for every chapter, an extensive review of what has been learned from physics education research, and guidelines for using active-learning techniques in your classroom.
- The **Instructor Solutions Manuals**, **Chapters 1–19** (ISBN 0-321-51621-4/978-0-321-51621-3) and **Chapters 20–43** (ISBN 0-321-51657-5/978-0-321-51657-2), written by the author and Professors Scott Nutter (Northern Kentucky University) and Larry Smith (Snow College), provide *complete* solutions to all the end-of-chapter problems. The solutions follow the four-step Model/Visualize/Solve/Assess

procedure used in the Problem-Solving Strategies and in all worked examples. The full text of each solution is available as an editable Word document and as a pdf file on the *Media Manager CD-ROMs* for your own use or for posting on your password-protected course website.

- The cross-platform **Media Manager CD-ROMs** (ISBN 0-321-51624-9/978-0-321-51624-4) provide invaluable and easy-to-use resources for your class, including jpg files of all the figures, photos, tables, key (boxed) equations, chapter summaries, and knowledge structures from the textbook. In addition, all Tactics Boxes, Problem-Solving Strategies, and key equations are provided in an editable Word format. Also included are Word versions and pdf files of the *Instructor Guide* and the *Instructor Solutions Manuals*, and complete *Student Workbook* answers as pdf files. PowerPoint Lecture Outlines and Classroom Response System "Clicker" Questions (including reading quizzes) formatted for PRS systems are provided as well. A simple browser interface allows you to quickly identify the material you need, add it to your shopping cart, and export for editing or directly into your PPT lectures. An **ActivPhysics CD-ROM,** providing a comprehensive library of more than 280 applets from *ActivPhysics OnLine,* and the **Computerized Test Bank CD-ROM** (see below) are also included.

- The online **Instructor Resource Center** (www.aw-bc.com/irc) provides updates to files on the Media Manager CD-ROMs. To obtain a login name and password, contact your Pearson Addison-Wesley sales representative.

- **MasteringPhysics**™ (www.masteringphysics.com) is the most widely used and educationally proven physics homework, tutorial, and assessment system available. It is designed to assign, assess, and track each student's progress using a wide diversity of extensively pre-tested problems. Icons throughout the book indicate that *MasteringPhysics*™ offers specific tutorials for all the textbook's Tactics Boxes and Problem-Solving Strategies, as well as all the end-of-chapter problems, Test Bank items, and Reading Quizzes. *MasteringPhysics*™ provides instructors with a fast and effective way to assign uncompromising, wide-ranging online homework assignments of just the right difficulty and duration. The powerful post-assignment diagnostics allow instructors to assess the progress of their class as a whole or to quickly identify individual students' areas of difficulty.

- **ActivPhysics OnLine**™ (accessed through the Self Study area within www.masteringphysics.com) provides a comprehensive library of more than 420 tried and tested *ActivPhysics* applets. In addition, it provides a suite of highly regarded applet-based tutorials developed by education pioneers Professors Alan Van Heuvelen and Paul D'Alessandris. The *ActivPhysics* icons that appear throughout the book direct students to specific interactive exercises that complement the textbook discussion.

  The online exercises are designed to encourage students to confront misconceptions, reason qualitatively about physical processes, experiment quantitatively, and learn to think critically. They cover all topics from mechanics to electricity and magnetism and from optics to modern physics. The highly acclaimed *ActivPhysics OnLine* companion workbooks help students work through complex concepts and understand them more clearly. More than 280 applets from the *ActivPhysics OnLine* library are also available on the Instructor *Media Manager CD-ROMs*.

- The **Printed Test Bank** (ISBN 0-321-51622-2/978-0-321-51622-0) and cross-platform **Computerized Test Bank** (included with the Media Manager CD-ROMs), prepared by Dr. Peter W. Murphy, contain more than 1500 high-quality problems, with a range of multiple-choice, true/false, short-answer, and regular homework-type questions. In the computerized version, more than half of the questions have numerical values that can be randomly assigned for each student.

- The **Transparency Acetates** (ISBN 0-321-51623-0/978-0-321-51623-7) provide more than 200 key figures from *Physics for Scientists and Engineers* for classroom presentation.

Student Supplements

- The **Student Solutions Manuals Chapters 1–19** (ISBN 0-321-51354-1/978-0-321-51356-0) and **Chapters 20–43** (ISBN 0-321-51356-8/978-0-321-51356-4), written by the author and Professors Scott Nutter (Northern Kentucky University) and Larry Smith (Snow College), provide *detailed* solutions to more than half of the odd-numbered end-of-chapter problems. The solutions follow the four-step Model/Visualize/Solve/Assess procedure used in the Problem-Solving Strategies and in all worked examples.

- **MasteringPhysics**™ (www.masteringphysics.com) is the most widely used and educationally proven physics homework, tutorial, and assessment system available. It is based on years of research into how students work physics problems and precisely where they need help. Studies show that students who use *Mastering-Physics*™ significantly increase their final scores compared to hand-written homework. *MasteringPhysics*™ achieves this improvement by providing students with instantaneous feedback specific to their wrong answers, simpler sub-problems upon request when they get stuck, and partial credit for their method(s) used. This individualized, 24/7 Socratic tutoring is recommended by nine out of ten students to their peers as the most effective and time-efficient way to study.

- **Pearson Tutor Services** (www.pearsontutorservices.com) Each student's subscription to MasteringPhysics also contains complimentary access to Pearson Tutor Services, powered by Smarthinking, Inc. By logging in with their Mastering-Physics ID and password, they will be connected to highly qualified e-structors™ who provide additional, interactive online tutoring on the major concepts of physics. Some restrictions apply; offer subject to change.

- **ActivPhysics OnLine**™ (accessed via www.masteringphysics.com) provides students with a suite of highly regarded applet-based self-study tutorials (see above). The *ActivPhysics* icons throughout the book direct students to specific exercises that complement the textbook discussion. The following workbooks provide a range of tutorial problems designed to use the *ActivPhysics OnLine* simulations, helping students work through complex concepts and understand them more clearly:

- **ActivPhysics OnLine Workbook Volume 1: Mechanics • Thermal Physics • Oscillations & Waves** (ISBN 0-8053-9060-X)

- **ActivPhysics OnLine Workbook Volume 2: Electricity & Magnetism • Optics • Modern Physics** (ISBN 0-8053-9061-8)

Acknowledgments

I have relied upon conversations with and, especially, the written publications of many members of the physics education community. Those who may recognize their influence include Arnold Arons, Uri Ganiel, Ibrahim Halloun, Richard Hake, Ken Heller, David Hestenes, Leonard Jossem, Jill Larkin, Priscilla Laws, John Mallinckrodt, Kandiah Manivannan, Lillian McDermott and members of the Physics Education Research Group at the University of Washington, David Meltzer, Edward "Joe" Redish, Fred Reif, Jeffery Saul, Rachel Scherr, Bruce Sherwood, Josip Slisko, David Sokoloff, Ronald Thornton, Sheila Tobias, and Alan Van Heuvelen. John Rigden, founder and director of the Introductory University Physics Project, provided the impetus that got me started down this path. Early development of the materials was supported by the National Science Foundation as the *Physics for the Year 2000* project; their support is gratefully acknowledged.

I am grateful to Larry Smith and Scott Nutter for the difficult task of writing the *Instructor Solutions Manuals*; to Jim Andrews and Rebecca Sabinovsky for writing the workbook answers; to Wayne Anderson, Jim Andrews, Dave Ettestad, Stuart

Field, Robert Glosser, and Charlie Hibbard for their contributions to the end-of-chapter problems; and to my colleague Matt Moelter for many valuable contributions and suggestions.

I especially want to thank my editor Adam Black, development editor Alice Houston, project editor Martha Steele, and all the other staff at Addison-Wesley for their enthusiasm and hard work on this project. Production supervisor Nancy Tabor, Jared Sterzer and the team at WestWords, Inc., and photo researcher Brian Donnelly get a good deal of the credit for making this complex project all come together. In addition to the reviewers and classroom testers listed below, who gave invaluable feedback, I am particularly grateful to Charlie Hibbard and Peter W. Murphy for their close scrutiny of every word and figure.

Finally, I am endlessly grateful to my wife Sally for her love, encouragement, and patience, and to our many cats (and especially to the memory of my faithful writing companion Spike) for their innate abilities to keep my keyboard and printer filled with cat fur and to always sit right in the middle of the carefully stacked page proofs.

Randy Knight, August 2007
rknight@calpoly.edu

Reviewers and Classroom Testers

Gary B. Adams, *Arizona State University*
Ed Adelson, *Ohio State University*
Kyle Altmann, *Elon University*
Wayne R. Anderson, *Sacramento City College*
James H. Andrews, *Youngstown State University*
Kevin Ankoviak, *Las Positas College*
David Balogh, *Fresno City College*
Dewayne Beery, *Buffalo State College*
Joseph Bellina, *Saint Mary's College*
James R. Benbrook, *University of Houston*
David Besson, *University of Kansas*
Randy Bohn, *University of Toledo*
Richard A. Bone, *Florida International University*
Gregory Boutis, *York College*
Art Braundmeier, *University of Southern Illinois, Edwardsville*
Carl Bromberg, *Michigan State University*
Meade Brooks, *Collin College*
Douglas Brown, *Cabrillo College*
Ronald Brown, *California Polytechnic State University, San Luis Obispo*
Mike Broyles, *Collin County Community College*
Debra Burris, *University of Central Arkansas*
James Carolan, *University of British Columbia*
Michael Chapman, *Georgia Tech University*
Norbert Chencinski, *College of Staten Island*
Kristi Concannon, *King's College*
Sean Cordry, *Northwestern College of Iowa*
Robert L. Corey, *South Dakota School of Mines*
Michael Crescimanno, *Youngstown State University*

Dennis Crossley, *University of Wisconsin–Sheboygan*
Wei Cui, *Purdue University*
Robert J. Culbertson, *Arizona State University*
Danielle Dalafave, *The College of New Jersey*
Purna C. Das, *Purdue University North Central*
Chad Davies, *Gordon College*
William DeGraffenreid, *California State University–Sacramento*
Dwain Desbien, *Estrella Mountain Community College*
John F. Devlin, *University of Michigan, Dearborn*
John DiBartolo, *Polytechnic University*
Alex Dickison, *Seminole Community College*
Chaden Djalali, *University of South Carolina*
Margaret Dobrowolska, *University of Notre Dame*
Sandra Doty, *Denison University*
Miles J. Dresser, *Washington State University*
Charlotte Elster, *Ohio University*
Robert J. Endorf, *University of Cincinnati*
Tilahun Eneyew, *Embry-Riddle Aeronautical University*
F. Paul Esposito, *University of Cincinnati*
John Evans, *Lee University*
Harold T. Evensen, *University of Wisconsin–Platteville*
Michael R. Falvo, *University of North Carolina*
Abbas Faridi, *Orange Coast College*
Nail Fazleev, *University of Texas–Arlington*
Stuart Field, *Colorado State University*
Daniel Finley, *University of New Mexico*
Jane D. Flood, *Muhlenberg College*
Michael Franklin, *Northwestern Michigan College*
Jonathan Friedman, *Amherst College*

Thomas Furtak, *Colorado School of Mines*
Alina Gabryszewska-Kukawa, *Delta State University*
Lev Gasparov, *University of North Florida*
Richard Gass, *University of Cincinnati*
J. David Gavenda, *University of Texas, Austin*
Stuart Gazes, *University of Chicago*
Katherine M. Gietzen, *Southwest Missouri State University*
Robert Glosser, *University of Texas, Dallas*
William Golightly, *University of California, Berkeley*
Paul Gresser, *University of Maryland*
C. Frank Griffin, *University of Akron*
John B. Gruber, *San Jose State University*
Stephen Haas, *University of Southern California*
John Hamilton, *University of Hawaii at Hilo*
Jason Harlow, *University of Toronto*
Randy Harris, *University of California, Davis*
Nathan Harshman, *American University*
J. E. Hasbun, *University of West Georgia*
Nicole Herbots, *Arizona State University*
Jim Hetrick, *University of Michigan–Dearborn*
Scott Hildreth, *Chabot College*
David Hobbs, *South Plains College*
Laurent Hodges, *Iowa State University*
Mark Hollabaugh, *Normandale Community College*
John L. Hubisz, *North Carolina State University*
Shane Hutson, *Vanderbilt University*
George Igo, *University of California, Los Angeles*
David C. Ingram, *Ohio University*
Bob Jacobsen, *University of California, Berkeley*
Rong-Sheng Jin, *Florida Institute of Technology*
Marty Johnston, *University of St. Thomas*
Stanley T. Jones, *University of Alabama*
Darrell Judge, *University of Southern California*
Pawan Kahol, *Missouri State University*
Teruki Kamon, *Texas A&M University*
Richard Karas, *California State University, San Marcos*
Deborah Katz, *U.S. Naval Academy*
Miron Kaufman, *Cleveland State University*
Katherine Keilty, *Kingwood College*
Roman Kezerashvili, *New York City College of Technology*
Peter Kjeer, *Bethany Lutheran College*
M. Kotlarchyk, *Rochester Institute of Technology*
Fred Krauss, *Delta College*
Cagliyan Kurdak, *University of Michigan*
Fred Kuttner, *University of California, Santa Cruz*
H. Sarma Lakkaraju, *San Jose State University*
Darrell R. Lamm, *Georgia Institute of Technology*
Robert LaMontagne, *Providence College*
Eric T. Lane, *University of Tennessee–Chattanooga*
Alessandra Lanzara, *University of California, Berkeley*
Lee H. LaRue, *Paris Junior College*
Sen-Ben Liao, *Massachusetts Institute of Technology*
Dean Livelybrooks, *University of Oregon*
Chun-Min Lo, *University of South Florida*
Olga Lobban, *Saint Mary's University*

Ramon Lopez, *Florida Institute of Technology*
Vaman M. Naik, *University of Michigan, Dearborn*
Kevin Mackay, *Grove City College*
Carl Maes, *University of Arizona*
Rizwan Mahmood, *Slippery Rock University*
Mani Manivannan, *Missouri State University*
Richard McCorkle, *University of Rhode Island*
James McDonald, *University of Hartford*
James McGuire, *Tulane University*
Stephen R. McNeil, *Brigham Young University–Idaho*
Theresa Moreau, *Amherst College*
Gary Morris, *Rice University*
Michael A. Morrison, *University of Oklahoma*
Richard Mowat, *North Carolina State University*
Eric Murray, *Georgia Institute of Technology*
Taha Mzoughi, *Mississippi State University*
Scott Nutter, *Northern Kentucky University*
Craig Ogilvie, *Iowa State University*
Benedict Y. Oh, *University of Wisconsin*
Martin Okafor, *Georgia Perimeter College*
Halina Opyrchal, *New Jersey Institute of Technology*
Yibin Pan, *University of Wisconsin-Madison*
Georgia Papaefthymiou, *Villanova University*
Peggy Perozzo, *Mary Baldwin College*
Brian K. Pickett, *Purdue University, Calumet*
Joe Pifer, *Rutgers University*
Dale Pleticha, *Gordon College*
Marie Plumb, *Jamestown Community College*
Robert Pompi, *SUNY-Binghamton*
David Potter, *Austin Community College–Rio Grande Campus*
Chandra Prayaga, *University of West Florida*
Didarul Qadir, *Central Michigan University*
Steve Quon, *Ventura College*
Michael Read, *College of the Siskiyous*
Lawrence Rees, *Brigham Young University*
Richard J. Reimann, *Boise State University*
Michael Rodman, *Spokane Falls Community College*
Sharon Rosell, *Central Washington University*
Anthony Russo, *Okaloosa-Walton Community College*
Freddie Salsbury, *Wake Forest University*
Otto F. Sankey, *Arizona State University*
Jeff Sanny, *Loyola Marymount University*
Rachel E. Scherr, *University of Maryland*
Carl Schneider, *U. S. Naval Academy*
Bruce Schumm, *University of California, Santa Cruz*
Bartlett M. Sheinberg, *Houston Community College*
Douglas Sherman, *San Jose State University*
Elizabeth H. Simmons, *Boston University*
Marlina Slamet, *Sacred Heart University*
Alan Slavin, *Trent College*
Larry Smith, *Snow College*
William S. Smith, *Boise State University*
Paul Sokol, *Pennsylvania State University*
LTC Bryndol Sones, *United States Military Academy*

Chris Sorensen, *Kansas State University*
Anna and Ivan Stern, *AW Tutor Center*
Gay B. Stewart, *University of Arkansas*
Michael Strauss, *University of Oklahoma*
Chin-Che Tin, *Auburn University*
Christos Valiotis, *Antelope Valley College*
Andrew Vanture, *Everett Community College*
Arthur Viescas, *Pennsylvania State University*
Ernst D. Von Meerwall, *University of Akron*
Chris Vuille, *Embry-Riddle Aeronautical University*
Jerry Wagner, *Rochester Institute of Technology*
Robert Webb, *Texas A&M University*

Zodiac Webster, *California State University, San Bernardino*
Robert Weidman, *Michigan Technical University*
Fred Weitfeldt, *Tulane University*
Jeff Allen Winger, *Mississippi State University*
Carey Witkov, *Broward Community College*
Ronald Zammit, *California Polytechnic State University, San Luis Obispo*
Darin T. Zimmerman, *Pennsylvania State University, Altoona*
Fredy Zypman, *Yeshiva University*

Preface to the Student

From Me to You

The most incomprehensible thing about the universe is that it is comprehensible.
  —Albert Einstein

The day I went into physics class it was death.
  —Sylvia Plath, *The Bell Jar*

Let's have a little chat before we start. A rather one-sided chat, admittedly, because you can't respond, but that's OK. I've talked with many of your fellow students over the years, so I have a pretty good idea of what's on your mind.

What's your reaction to taking physics? Fear and loathing? Uncertainty? Excitement? All of the above? Let's face it, physics has a bit of an image problem on campus. You've probably heard that it's difficult, maybe downright impossible unless you're an Einstein. Things that you've heard, your experiences in other science courses, and many other factors all color your *expectations* about what this course is going to be like.

It's true that there are many new ideas to be learned in physics and that the course, like college courses in general, is going to be much faster paced than science courses you had in high school. I think it's fair to say that it will be an *intense* course. But we can avoid many potential problems and difficulties if we can establish, here at the beginning, what this course is about and what is expected of you—and of me!

Just what is physics, anyway? Physics is a way of thinking about the physical aspects of nature. Physics is not better than art or biology or poetry or religion, which are also ways to think about nature; it's simply different. One of the things this course will emphasize is that physics is a human endeavor. The ideas presented in this book were not found in a cave or conveyed to us by aliens; they were discovered and developed by real people engaged in a struggle with real issues. I hope to convey to you something of the history and the process by which we have come to accept the principles that form the foundation of today's science and engineering.

You might be surprised to hear that physics is not about "facts." Oh, not that facts are unimportant, but physics is far more focused on discovering *relationships* that exist between facts and *patterns* that exist in nature than on learning facts for their own sake. As a consequence, there's not a lot of memorization when you study physics. Some—there are still definitions and equations to learn—but less than in many other courses. Our emphasis, instead, will be on thinking and reasoning. This is important to factor into your expectations for the course.

Perhaps most important of all, *physics is not math!* Physics is much broader. We're going to look for patterns and relationships in nature, develop the logic that relates different ideas, and search for the reasons *why* things happen as they do. In doing so, we're going to stress qualitative reasoning, pictorial and graphical reasoning, and reasoning by analogy. And yes, we will use math, but it's just one tool among many.

It will save you much frustration if you're aware of this physics–math distinction up front. Many of you, I know, want to find a formula and plug numbers into it—

(a) X-ray diffraction pattern

(b) Electron diffraction pattern

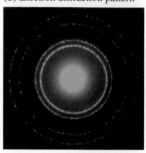

that is, to do a math problem. Maybe that worked in high school science courses, but it is *not* what this course expects of you. We'll certainly do many calculations, but the specific numbers are usually the last and least important step in the analysis.

Physics is about recognizing patterns. For example, the top photograph is an x-ray diffraction pattern showing how a focused beam of x rays spreads out after passing through a crystal. The bottom photograph shows what happens when a focused beam of electrons is shot through the same crystal. What does the obvious similarity in these two photographs tell us about the nature of light and the nature of matter?

As you study, you'll sometimes be baffled, puzzled, and confused. That's perfectly normal and to be expected. Making mistakes is OK too *if* you're willing to learn from the experience. No one is born knowing how to do physics any more than he or she is born knowing how to play the piano or shoot basketballs. The ability to do physics comes from practice, repetition, and struggling with the ideas until you "own" them and can apply them yourself in new situations. There's no way to make learning effortless, at least for anything worth learning, so expect to have some difficult moments ahead. But also expect to have some moments of excitement at the joy of discovery. There will be instants at which the pieces suddenly click into place and you *know* that you understand a powerful idea. There will be times when you'll surprise yourself by successfully working a difficult problem that you didn't think you could solve. My hope, as an author, is that the excitement and sense of adventure will far outweigh the difficulties and frustrations.

Getting the Most Out of Your Course

Many of you, I suspect, would like to know the "best" way to study for this course. There is no best way. People are different, and what works for one student is less effective for another. But I do want to stress that *reading the text* is vitally important. Class time will be used to clarify difficulties and to develop tools for using the knowledge, but your instructor will *not* use class time simply to repeat information in the text. The basic knowledge for this course is written down on these pages, and the *number-one expectation* is that you will read carefully and thoroughly to find and learn that knowledge.

Despite there being no best way to study, I will suggest *one* way that is successful for many students. It consists of the following four steps:

1. **Read each chapter *before* it is discussed in class.** I cannot stress too strongly how important this step is. Class attendance is much more effective if you are prepared. When you first read a chapter, focus on learning new vocabulary, definitions, and notation. There's a list of terms and notations at the end of each chapter. Learn them! You won't understand what's being discussed or how the ideas are being used if you don't know what the terms and symbols mean.

2. **Participate actively in class.** Take notes, ask and answer questions, and participate in discussion groups. There is ample scientific evidence that *active participation* is much more effective for learning science than passive listening.

3. **After class, go back for a careful re-reading of the chapter.** In your second reading, pay closer attention to the details and the worked examples. Look for the *logic* behind each example (I've highlighted this to make it clear), not just at what formula is being used. Do the *Student Workbook* exercises for each section as you finish your reading of it.

4. **Finally, apply what you have learned to the homework problems at the end of each chapter.** I strongly encourage you to form a study group with two or three classmates. There's good evidence that students who study regularly with a group do better than the rugged individualists who try to go it alone.

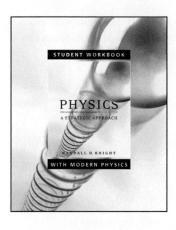
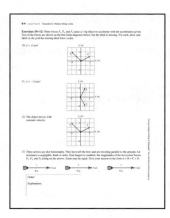

Did someone mention a workbook? The companion *Student Workbook* is a vital part of the course. Its questions and exercises ask you to reason *qualitatively,* to use graphical information, and to give explanations. It is through these exercises that you will learn what the concepts mean and will practice the reasoning skills appropriate to the chapter. You will then have acquired the baseline knowledge and confidence you need *before* turning to the end-of-chapter homework problems. In sports or in music, you would never think of performing before you practice, so why would you want to do so in physics? The workbook is where you practice and work on basic skills.

Many of you, I know, will be tempted to go straight to the homework problems and then thumb through the text looking for a formula that seems like it will work. That approach will not succeed in this course, and it's guaranteed to make you frustrated and discouraged. Very few homework problems are of the "plug and chug" variety where you simply put numbers into a formula. To work the homework problems successfully, you need a better study strategy—either the one outlined above or your own—that helps you learn the concepts and the relationships between the ideas.

A traditional guideline in college is to study two hours outside of class for every hour spent in class, and this text is designed with that expectation. Of course, two hours is an average. Some chapters are fairly straightforward and will go quickly. Others likely will require much more than two study hours per class hour.

Getting the Most Out of Your Textbook

Your textbook provides many features designed to help you learn the concepts of physics and solve problems more effectively.

- **TACTICS BOXES** give step-by-step procedures for particular skills, such as interpreting graphs or drawing special diagrams. Tactics Box steps are explicitly illustrated in subsequent worked examples, and these are often the starting point of a full *Problem-Solving Strategy.*

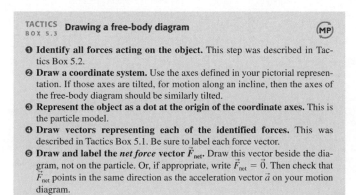
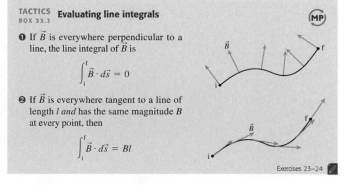

TACTICS BOX 5.3 **Drawing a free-body diagram** (MP)

❶ **Identify all forces acting on the object.** This step was described in Tactics Box 5.2.
❷ **Draw a coordinate system.** Use the axes defined in your pictorial representation. If those axes are tilted, for motion along an incline, then the axes of the free-body diagram should be similarly tilted.
❸ **Represent the object as a dot at the origin of the coordinate axes.** This is the particle model.
❹ **Draw vectors representing each of the identified forces.** This was described in Tactics Box 5.1. Be sure to label each force vector.
❺ **Draw and label the *net force* vector $\vec{F}_{net}$.** Draw this vector beside the diagram, not on the particle. Or, if appropriate, write $\vec{F}_{net} = \vec{0}$. Then check that $\vec{F}_{net}$ points in the same direction as the acceleration vector $\vec{a}$ on your motion diagram.

Exercises 24–29

TACTICS BOX 33.3 **Evaluating line integrals** (MP)

❶ If $\vec{B}$ is everywhere perpendicular to a line, the line integral of $\vec{B}$ is
$$\int_i^f \vec{B} \cdot d\vec{s} = 0$$

❷ If $\vec{B}$ is everywhere tangent to a line of length l *and* has the same magnitude B at every point, then
$$\int_i^f \vec{B} \cdot d\vec{s} = Bl$$

Exercises 23–24

■ **PROBLEM-SOLVING STRATEGIES** are provided for each broad class of problems—problems characteristic of a chapter or group of chapters. The strategies follow a consistent four-step approach to help you develop confidence and proficient problem-solving skills: **MODEL, VISUALIZE, SOLVE, ASSESS.**

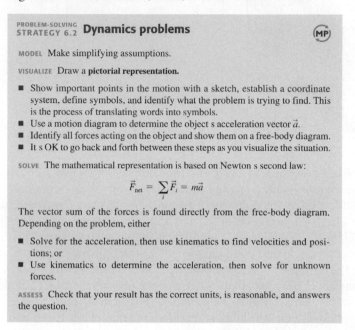

PROBLEM-SOLVING **STRATEGY 6.2** **Dynamics problems**   (MP)

MODEL Make simplifying assumptions.

VISUALIZE Draw a **pictorial representation.**

■ Show important points in the motion with a sketch, establish a coordinate system, define symbols, and identify what the problem is trying to find. This is the process of translating words into symbols.
■ Use a motion s diagram to determine the object s acceleration vector $\vec{a}$.
■ Identify all forces acting on the object and show them on a free-body diagram.
■ It s OK to go back and forth between these steps as you visualize the situation.

SOLVE The mathematical representation is based on Newton s second law:

$$\vec{F}_{net} = \sum_i \vec{F}_i = m\vec{a}$$

The vector sum of the forces is found directly from the free-body diagram. Depending on the problem, either

■ Solve for the acceleration, then use kinematics to find velocities and positions; or
■ Use kinematics to determine the acceleration, then solve for unknown forces.

ASSESS Check that your result has the correct units, is reasonable, and answers the question.

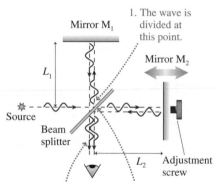

Annotated **FIGURE** showing the operation of the Michelson interferometer.

■ Worked **EXAMPLES** illustrate good problem-solving practices through the consistent use of the four-step problem-solving approach and, where appropriate, the Tactics Box steps. The worked examples are often very detailed and carefully lead you through the *reasoning* behind the solution as well as the numerical calculations. A careful study of the reasoning will help you apply the concepts and techniques to the new and novel problems you will encounter in homework assignments and on exams.

■ **NOTE ▶** paragraphs alert you to common mistakes and point out useful tips for tackling problems.

■ **STOP TO THINK** questions embedded in the chapter allow you to quickly assess whether you've understood the main idea of a section. A correct answer will give you confidence to move on to the next section. An incorrect answer will alert you to re-read the previous section.

■ Blue annotations on figures help you better understand what the figure is showing. They will help you to interpret graphs; translate between graphs, math, and pictures; grasp difficult concepts through a visual analogy; and develop many other important skills.

■ *Pencil sketches* provide practical examples of the figures you should draw yourself when solving a problem.

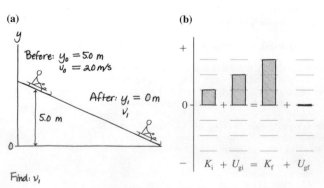

Pencil-sketch **FIGURE** showing a toboggan going down a hill and its energy bar chart.

- The learning goals and links that begin each chapter outline what to focus on in the chapter ahead and what you need to remember from previous chapters.
  - ▶ **Looking Ahead** lists key concepts and skills you will learn in the coming chapter.
  - ◀ **Looking Back** highlights important topics you should review from previous chapters.
- Schematic *Chapter Summaries* help you organize what you have learned into a hierarchy, from general principles (top) to applications (bottom). Side-by-side pictorial, graphical, textual, and mathematical representations are used to help you translate between these key representations.
- *Part Overviews and Summaries* provide a global framework for what you are learning. Each part begins with an overview of the chapters ahead and concludes with a broad summary to help you to connect the concepts presented in that set of chapters. KNOWLEDGE STRUCTURE tables in the Part Summaries, similar to the Chapter Summaries, help you to see the forest rather than just the trees.

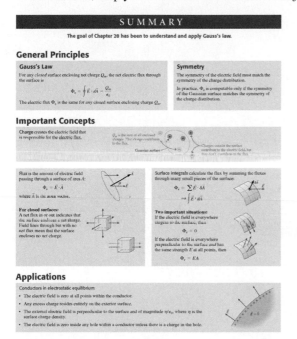

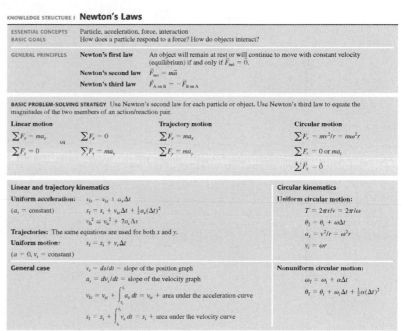

Now that you know more about what is expected of you, what can you expect of me? That's a little trickier because the book is already written! Nonetheless, the book was prepared on the basis of what I think my students throughout the years have expected—and wanted—from their physics textbook. Further, I've listened to the extensive feedback I have received from thousands of students like you, and their instructors, who used the first edition of this book.

You should know that these course materials—the text and the workbook—are based on extensive research about how students learn physics and the challenges they face. The effectiveness of many of the exercises has been demonstrated through extensive class testing. I've written the book in an informal style that I hope you will find appealing and that will encourage you to do the reading. And, finally, I have endeavored to make clear not only that physics, as a technical body of knowledge, is relevant to your profession but also that physics is an exciting adventure of the human mind.

I hope you'll enjoy the time we're going to spend together.

Detailed Contents

Volume 1 contains chapters 1–15; Volume 2 contains chapters 16–19; Volume 3 contains chapters 20–25; Volume 4 contains chapters 26–37; Volume 5 contains chapters 37–43.

Thermodynamics

A modern jet engine is a marvel of technical ingenuity. Understanding how a jet engine works requires understanding the thermodynamics of gases and heat engines.

OVERVIEW

It's All About Energy

Thermodynamics—the science of energy in its broadest context—arose hand in hand with the industrial revolution as the systematic study of converting heat energy into mechanical motion and work. Hence the name *thermo + dynamics*. Indeed, the analysis of engines and generators of various kinds remains the focus of engineering thermodynamics. But thermodynamics, as a science, now extends to all forms of energy conversions, including those involving living organisms. For example:

- **Engines** convert the energy of a fuel into the mechanical energy of moving pistons, gears, and wheels.
- **Fuel cells** convert chemical energy into electrical energy.
- **Photovoltaic cells** convert the electromagnetic energy of light into electrical energy.
- **Lasers** convert electrical energy into the electromagnetic energy of light.
- **Organisms** convert the chemical energy of food into a variety of other forms of energy, including kinetic energy, sound energy, and thermal energy.

The major goals of Part IV are to understand both *how* energy transformations such as these take place and *how efficient* they are. We'll discover that the laws of thermodynamics place limits on the efficiency of energy transformations, and understanding these limits is essential for analyzing the very real energy needs of society in the 21st century.

Our ultimate destination in Part IV is an understanding of the thermodynamics of *heat engines*. A heat engine is a device, such as a power plant or an internal combustion engine, that transforms heat energy into useful work. These are the devices that power our modern society.

Understanding how to transform heat into work will be a significant achievement, but we first have many steps to take along the way. We need to understand the concepts of temperature and pressure. We need to learn about the properties of solids, liquids, and gases. Most important, we need to expand our view of energy to include *heat,* the energy that is transferred between two systems at different temperatures.

At a deeper level, we need to see how these concepts are connected to the underlying microphysics of randomly moving molecules. We will find that the familiar concepts of thermodynamics, such as temperature and pressure, have their roots in atomic-level motion and collisions. We will also find it possible to learn a great deal about the properties of molecules, such as their speeds, on the basis of purely macroscopic measurements. This *micro/macro connection* will lead to the second law of thermodynamics, one of the most subtle but also one of the most profound and far-reaching statements in physics.

Only after all these steps have been taken will we be able to analyze a real heat engine. It is an ambitious goal, but one we can achieve.

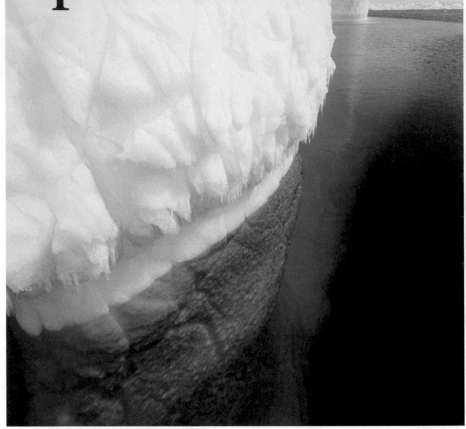

16 A Macroscopic Description of Matter

Solid, liquid, and gas—the three phases of matter.

▶ **Looking Ahead**

The goal of Chapter 16 is to learn the characteristics of macroscopic systems. In this chapter you will learn to:

- Understand the basic properties of solids, liquids, and gases.
- Interpret a phase diagram.
- Work with different temperature scales.
- Use the ideal-gas law.
- Understand ideal-gas processes and represent them on a pV diagram.

◀ **Looking Back**

The material in this chapter depends on thermal energy and the properties of fluids. Please review:

- Section 11.7 Thermal energy.
- Sections 15.1–15.3 Fluids and pressure.

A room full of air, a beaker of water, and this floating iceberg are examples of macroscopic systems, systems large enough to see or touch. These are the systems of our everyday experience. Our goal in this chapter is twofold:

- To learn what kind of physical properties characterize macroscopic systems.
- To begin the process of connecting a system's macroscopic properties to the underlying motions of the atoms in the system.

The properties of a macroscopic system as a whole are called its **bulk properties.** One fairly obvious example is the system's mass. Other bulk properties are volume, density, temperature, and pressure. Macroscopic systems are also characterized as being either solid, liquid, or gas. These are called the *phases* of matter, and we'll be interested in when and how a system changes from one phase to another.

Ultimately we would like to understand the macroscopic properties of solids, liquids, and gases in terms of the microscopic motions of their atoms and molecules. Developing this **micro/macro connection** will take several chapters, but we'll start laying the foundations in this chapter. This effort to understand macroscopic properties in terms of particle-like atoms will pay handsome dividends when we later come to electricity and then quantum physics.

16.1 Solids, Liquids, and Gases

The ice cube you take out of the freezer soon becomes a puddle of liquid water. Then, more slowly, it evaporates to become water vapor in the air. Water is unique. It is the only substance whose three **phases**—solid, liquid, and gas—are familiar from every-day experience.

Each of the elements and most compounds can exist as a solid, liquid, or gas. The change between liquid and solid (freezing or melting) or between liquid and gas (boiling or condensing) is called a **phase change.** We're familiar with only one, or perhaps two, of the phases of most substances because their melting point and/or boiling point are far outside the range of normal human experience.

The notion of three distinct phases is less useful for more complex systems. A piece of wood is solid, but liquid wood and gaseous wood don't exist. *Liquid crystals,* which are used to display the numbers on your digital watch, have characteristics of both solids *and* liquids. Complex systems have many interesting properties, but this text will focus on macroscopic systems for which the three phases are distinct.

NOTE ▶ This use of the word "phase" has no relationship at all to the *phase* or *phase constant* of simple harmonic motion and waves. ◀

Metals as hard as steel can be melted and, at a high enough temperature, even boiled.

Solids, liquids, and gases

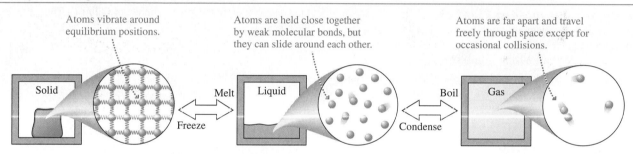

Atoms vibrate around equilibrium positions.

Atoms are held close together by weak molecular bonds, but they can slide around each other.

Atoms are far apart and travel freely through space except for occasional collisions.

Solid ⟷ Melt / Freeze ⟷ Liquid ⟷ Boil / Condense ⟷ Gas

A **solid** is a rigid macroscopic system with a definite shape and volume. It consists of particle-like atoms connected together by spring-like molecular bonds. Each atom vibrates around an equilibrium position, but an atom is *not* free to move inside the solid. Solids are nearly *incompressible,* telling us that the atoms in a solid are just about as close together as they can get.

The solid shown here is a **crystal,** meaning that the atoms are arranged in a periodic array. The elements and many compounds have a crystal structure when in their solid phase. In other solids, such as glass, the atoms are frozen into random positions. These are called **amorphous solids.**

A **liquid** is more complicated than either a solid or a gas. Like a solid, a liquid is nearly *incompressible.* This tells us that the molecules in a liquid are about as close together as they can get. Like a gas, a liquid flows and deforms to fit the shape of its container. The fluid nature of a liquid tells us that the molecules are free to move around.

Together, these observations suggest a model in which the molecules of the liquid are loosely held together by weak molecular bonds. The bonds are strong enough that the molecules never get far apart but not strong enough to prevent the molecules from sliding around each other.

A **gas** is a system in which each molecule moves through space as a free, noninteracting particle until, on occasion, it collides with another molecule or with the wall of the container. A gas is a *fluid.* A gas is also highly *compressible,* telling us that there is lots of space between the molecules.

Gases are fairly simple macroscopic systems; hence many of our examples in Part IV will be based on gases.

State Variables

The parameters used to characterize or describe a macroscopic system are known as **state variables** because, taken all together, they describe the *state* of the macroscopic system. You met some state variables in earlier chapters: volume, pressure, mass, mass density, and thermal energy. We'll soon introduce several new state variables: moles, number density, and, most important, the temperature T.

The state variables are not all independent of each other. For example, you learned in Chapter 15 that a system's mass density ρ is defined in terms of the system's mass M and volume V as

$$\rho = \frac{M}{V} \quad \text{(mass density)} \tag{16.1}$$

TABLE 16.1 Densities of materials

Substance	ρ (kg/m^3)
Air at STP*	1.3
Ethyl alcohol	790
Water (solid)	920
Water (liquid)	1000
Aluminum	2700
Copper	8920
Gold	19,300
Iron	7870
Lead	11,300
Mercury	13,600
Silicon	2330

*$T = 0°C$, $p = 1$ atm

In this chapter we'll use an uppercase M for the system mass and a lowercase m for the mass of an atom. Table 16.1 is a short list of mass densities.

If we change the value of any of the state variables, then we change the state of the system. For example, to *compress* a gas means to decrease its volume. Other state variables, such as pressure and temperature, may also change as the volume changes. The symbol Δ represents a *change* in the value of a state variable. That is, ΔT is a *change* of temperature and Δp is a *change* of pressure. **For any quantity X, ΔX is always $X_f - X_i$, the final value minus the initial value.**

A system is said to be in **thermal equilibrium** if its state variables are constant and not changing. As an example, a gas is in thermal equilibrium if it has been left undisturbed long enough for p, V, and T to reach steady values. One of the important goals of Part IV is to establish the conditions under which a macroscopic system reaches thermal equilibrium.

EXAMPLE 16.1 The mass of a lead pipe

A project on which you are working uses a cylindrical lead pipe with outer and inner diameters of 4.0 cm and 3.5 cm, respectively, and a length of 50 cm. What is its mass?

SOLVE The mass density of lead is $\rho_{lead} = 11,300$ kg/m³. The volume of a circular cylinder of length l is $V = \pi r^2 l$. In this case we need to find the volume of the outer cylinder, of radius r_2, *minus*

the volume of air in the inner cylinder, of radius r_1. The volume of the pipe is

$$V = \pi r_2^2 l - \pi r_1^2 l = \pi(r_2^2 - r_1^2)l = 1.47 \times 10^{-4} \text{ m}^3$$

Hence the pipe's mass is

$$M = \rho_{lead} V = 1.7 \text{ kg}$$

STOP TO THINK 16.1   The pressure in a system is measured to be 60 kPa. At a later time the pressure is 40 kPa. The value of Δp is

a. 60 kPa   b. 40 kPa   c. 20 kPa   d. -20 kPa

16.2 Atoms and Moles

The mass of a macroscopic system is directly related to the total number of atoms or molecules in the system, denoted N. Because N is determined simply by counting, it is a number with no units. A typical macroscopic system has $N \sim 10^{25}$ atoms, an incredibly large number.

The symbol $\sim$, if you are not familiar with it, stands for "has the order of magnitude." It means that the number is known only to within a factor of 10 or so. The statement $N \sim 10^{25}$, which is read "N is of order 10^{25}," implies that N is somewhere in the range 10^{24} to 10^{26}. It is far less precise than the "approximately equal" symbol $\approx$. As we begin to deal with large numbers it will often be necessary to distinguish "really large" numbers, such as 10^{25}, from "small" numbers such as a mere 10^5. Saying $N \sim 10^{25}$ gives us a rough idea of how large N is and allows us to know that it differs significantly from 10^5 or even 10^{15}.

It is often useful to know the number of atoms or molecules per cubic meter in a system. We call this quantity the **number density.** It characterizes how densely the atoms are packed together within the system. In an N-atom system that fills volume V, the number density is

$$\frac{N}{V} \quad \text{(number density)} \qquad (16.2)$$

The SI units of number density are m^{-3}. The number density of atoms in a solid is $(N/V)_{\text{solid}} \sim 10^{29} \, m^{-3}$. The number density of a gas depends on the pressure, but is usually less than $10^{27} \, m^{-3}$. As **FIGURE 16.1** shows, **the value of N/V in a *uniform* system is independent of the volume V.** That is, the number density is the same whether you look at the whole system or just a portion of it.

NOTE ▶ While we might say "There are 100 tennis balls per cubic meter," or "There are 10^{29} atoms per cubic meter," tennis balls and atoms are not units. The units of N/V are simply m^{-3}. ◀

Atomic Mass and Atomic Mass Number

You will recall from chemistry that atoms of different elements have different masses. The mass of an atom is determined primarily by its most massive constituents, the protons and neutrons in its nucleus. The *sum* of the number of protons and neutrons is called the **atomic mass number** A:

$$A = \text{number of protons} + \text{number of neutrons}$$

A, which by definition is an integer, is written as a leading superscript on the atomic symbol. For example, the common isotope of hydrogen, with one proton and no neutrons, is ^1H. The "heavy hydrogen" isotope called *deuterium,* which includes one neutron, is ^2H. The primary isotope of carbon, with six protons (which makes it carbon) and six neutrons, is ^{12}C. The radioactive isotope ^{14}C, used for carbon dating of archeological finds, contains six protons and eight neutrons.

The **atomic mass** scale is established by defining the mass of ^{12}C to be exactly 12 u, where u is the symbol for the **atomic mass unit.** That is, $m(^{12}C) = 12$ u. The atomic mass of any other atom is its mass relative to ^{12}C. For example, careful experiments with hydrogen find that the mass *ratio* $m(^1H)/m(^{12}C)$ is 1.0078/12. Thus the atomic mass of hydrogen is $m(^1H) = 1.0078$ u.

The numerical value of the atomic mass of ^1H is close to, but not exactly, its atomic mass number $A = 1$. The slight difference is due to the electron mass and to various relativistic effects. For our purposes, it will be sufficient to overlook the slight difference and use the integer atomic mass numbers as the values of the atomic mass. That is, we'll use $m(^1H) = 1$ u, $m(^4He) = 4$ u, and $m(^{16}O) = 16$ u. For molecules, the **molecular mass** is the sum of the atomic masses of the atoms forming the molecule. Thus the molecular mass of the diatomic molecule O_2, the constituent of oxygen gas, is $m(O_2) = 32$ u.

NOTE ▶ An element's atomic mass number is *not* the same as its atomic number. The *atomic number,* the element's position in the periodic table, is the number of protons in the nucleus. ◀

Table 16.2 shows the atomic mass numbers of some of the elements that we'll use for examples and homework problems. A complete periodic table of the elements, including atomic masses, is found in Appendix B.

Moles and Molar Mass

One way to specify the amount of substance in a macroscopic system is to give its mass. Another way, one connected to the number of atoms, is to measure the amount of substance in *moles.* By definition, one **mole** of matter, be it solid, liquid, or gas, is

FIGURE 16.1 The number density of a uniform system is independent of the volume.

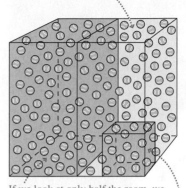

A 100 m^3 room has 10,000 tennis balls bouncing around. The number density of tennis balls in the room is $N/V = 10,000/100 \, m^3 = 100 \, m^{-3}$.

If we look at only half the room, we would find 5000 balls in 50 m^3, again giving $N/V = 5000/50 \, m^3 = 100 \, m^{-3}$.

In one-tenth of the room, we would find 1000 balls in 10 m^3, again giving $N/V = 1000/10 \, m^3 = 100 \, m^{-3}$.

TABLE 16.2 Some atomic mass numbers

Element		A
^1H	Hydrogen	1
^4He	Helium	4
^{12}C	Carbon	12
^{14}N	Nitrogen	14
^{16}O	Oxygen	16
^{20}Ne	Neon	20
^{27}Al	Aluminum	27
^{40}Ar	Argon	40
^{207}Pb	Lead	207

One mole of helium, sulfur, copper, and mercury.

TABLE 16.3 Monatomic and diatomic gases

Monatomic		Diatomic	
He	Helium	H_2	Hydrogen
Ne	Neon	N_2	Nitrogen
Ar	Argon	O_2	Oxygen

the amount of substance containing as many basic particles as there are atoms in 12 g of ^{12}C. Many decades of ingenious experiments have determined that there are 6.02×10^{23} atoms in 12 g of ^{12}C, so we can say that 1 mole of substance, abbreviated 1 mol, is 6.02×10^{23} basic particles.

The basic particle depends on the substance. Helium is a **monatomic gas,** meaning that the basic particle is the helium atom. Thus 6.02×10^{23} helium atoms are 1 mol of helium. But oxygen gas is a **diatomic gas** because the basic particle is the two-atom diatomic molecule O_2. 1 mol of oxygen gas contains 6.02×10^{23} *molecules* of O_2 and thus $2 \times 6.02 \times 10^{23}$ oxygen atoms. Table 16.3 lists the monatomic and diatomic gases that we will use for examples and homework problems.

The number of basic particles per mole of substance is called **Avogadro's number,** N_A. The value of Avogadro's number is

$$N_A = 6.02 \times 10^{23} \text{ mol}^{-1}$$

Avogadro's number, like the gravitational constant G, is one of the basic constants of nature.

Despite its name, Avogadro's number is not simply "a number"; it has units. Because there are N_A particles per mole, the number of moles in a substance containing N basic particles is

$$n = \frac{N}{N_A} \tag{16.3}$$

where n is the symbol for moles.

Avogadro's number allows us to determine atomic masses in kilograms. Knowing that N_A ^{12}C atoms have a mass of 12 g, we know the mass of one ^{12}C atom must be

$$m(^{12}C) = \frac{12 \text{ g}}{6.02 \times 10^{23}} = 1.993 \times 10^{-23} \text{ g} = 1.993 \times 10^{-26} \text{ kg}$$

We defined the atomic mass scale such that $m(^{12}C) = 12$ u. Thus the conversion factor between atomic mass units and kilograms is

$$1 \text{ u} = \frac{m(^{12}C)}{12} = 1.66 \times 10^{-27} \text{ kg}$$

This conversion factor allows us to calculate the mass in kg of any atom. For example, a ^{20}Ne atom has atomic mass $m(^{20}Ne) = 20$ u. Multiplying by 1.66×10^{-27} kg/u gives $m(^{20}Ne) = 3.32 \times 10^{-26}$ kg. If the atomic mass is specified in kilograms, the number of atoms in a system of mass M can be found from

$$N = \frac{M}{m} \tag{16.4}$$

The **molar mass** of a substance is the mass *in grams* of 1 mol of substance. The molar mass, which we'll designate M_{mol}, has units g/mol. By definition, the molar mass of ^{12}C is 12 g/mol. For other substances, whose atomic or molecular masses are given relative to ^{12}C, the numerical value of the molar mass equals the numerical value of the atomic or molecular mass. For example, the molar mass of He, with $m = 4$ u, is $M_{mol}(He) = 4$ g/mol and the molar mass of diatomic O_2 is $M_{mol}(O_2) = 32$ g/mol.

Equation 16.4 uses the atomic mass to find the number of atoms in a system. Similarly, you can use the molar mass to determine the number of moles. For a system of mass M consisting of atoms or molecules with molar mass M_{mol},

$$n = \frac{M \text{ (in grams)}}{M_{mol}} \tag{16.5}$$

NOTE ▶ Equation 16.5 is the one of the few instances where the proper units are *grams* rather than kilograms. ◀

EXAMPLE 16.2 Moles of oxygen

100 g of oxygen gas is how many moles of oxygen?

SOLVE We can do the calculation two ways. First, let's determine the number of molecules in 100 g of oxygen. The diatomic oxygen molecule O_2 has molecular mass $m = 32$ u. Converting this to kg, we get the mass of one molecule:

$$m = 32 \text{ u} \times \frac{1.66 \times 10^{-27} \text{ kg}}{1 \text{ u}} = 5.31 \times 10^{-26} \text{ kg}$$

Thus the number of molecules in 100 g = 0.10 kg is

$$N = \frac{M}{m} = \frac{0.100 \text{ kg}}{5.31 \times 10^{-26} \text{ kg}} = 1.88 \times 10^{24}$$

Knowing the number of molecules gives us the number of moles:

$$n = \frac{N}{N_A} = 3.13 \text{ mol}$$

Alternatively, we can use Equation 16.5 to find

$$n = \frac{M \text{ (in grams)}}{M_{mol}} = \frac{100 \text{ g}}{32 \text{ g/mol}} = 3.13 \text{ mol}$$

STOP TO THINK 16.2 Which system contains more atoms: 5 mol of helium ($A = 4$) or 1 mol of neon ($A = 20$)?

a. Helium.　　　b. Neon.　　　c. They have the same number of atoms.

16.3 Temperature

We are all familiar with the idea of temperature. You hear the word used nearly every day. But just what is temperature a measure of? Mass is a measure of the amount of substance in a system. Velocity is a measure of how fast a system moves. What physical property of the system have you determined if you measure its temperature?

We will begin with the commonsense idea that temperature is a measure of how "hot" or "cold" a system is. These are properties that we can judge without needing an elaborate theory. As we develop these ideas, we'll find that **temperature** T is related to a system's *thermal energy*. We defined thermal energy in Chapter 10 as the kinetic and potential energy of the atoms and molecules in a system as they vibrate (a solid) or move around (a gas). A system has more thermal energy when it is "hot" than when it is "cold." We'll study temperature more carefully in Chapter 18 and replace these vague notions of hot and cold with a precise relationship between temperature and thermal energy.

To start, we need a means to measure the temperature of a system. This is what a *thermometer* does. A thermometer can be any small macroscopic system that undergoes a measurable change as it exchanges thermal energy with its surroundings. It is placed in contact with a larger system whose temperature it will measure. In a common glass-tube thermometer, for example, a small volume of mercury or alcohol expands or contracts when placed in contact with a "hot" or "cold" object. The object's temperature is determined by the length of the column of liquid.

Other thermometers include:

- Bimetallic strips (two strips of different metals sandwiched together) that curl and uncurl as the temperature changes. These are used in thermostats, such as the one in your house.
- Thermocouples that generate a small voltage depending on the temperature. Thermocouples are widely used for sensing temperatures in inhospitable environments, such as in your car's engine.
- Ideal gases, whose pressure varies with the temperature. We will look at an example of a gas thermometer in a minute.

A thermometer needs a *temperature scale* to be a useful measuring device. In 1742, the Swedish astronomer Anders Celsius sealed mercury into a small capillary tube and observed how it moved up and down the tube as the temperature changed. He selected two temperatures that anyone could reproduce, the freezing and boiling points of pure water, and labeled them 0 and 100. He then marked off the glass tube into one hundred

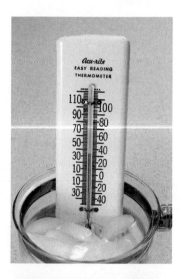

Thermal expansion of the liquid in the thermometer tube pushes it higher in the hot water than in the ice water.

equal intervals between these two reference points. By doing so, he invented the temperature scale that we today call the *Celsius scale*. The units of the Celsius temperature scale are "degrees Celsius," which we abbreviate °C. Note that the degree symbol ° is part of the unit, not part of the number.

NOTE ▶ Because of the 100 equal intervals, the Celsius scale is also called the *centigrade scale.* ◀

The *Fahrenheit scale,* still widely used in the United States, is related to the Celsius scale by

$$T_F = \frac{9}{5}T_C + 32° \tag{16.6}$$

Table 16.4 lists several temperatures measured on the Celsius and Fahrenheit scales and also on the Kelvin scale.

TABLE 16.4 Temperatures measured with different scales

Temperature	$T\,(°C)$	$T\,(K)$	$T\,(°F)$
Melting point of iron	1538	1811	2800
Boiling point of water	100	373	212
Normal body temperature	37.0	310	98.6
Room temperature	20	293	68
Freezing point of water	0	273	32
Boiling point of nitrogen	−196	77	−321
Absolute zero	−273	0	−460

Absolute Zero and Absolute Temperature

Any physical property that changes with temperature can be used as a thermometer. In practice, the most useful thermometers have a physical property that changes *linearly* with temperature. One of the most important scientific thermometers is the **constant-volume gas thermometer** shown in **FIGURE 16.2a.** This thermometer depends on the fact that the *absolute* pressure (not the gauge pressure) of a gas in a sealed container increases linearly as the temperature increases.

A gas thermometer is first calibrated by recording the pressure at two reference temperatures, such as the boiling and freezing points of water. These two points are plotted on a pressure-versus-temperature graph and a straight line is drawn through them. The gas bulb is then brought into contact with the system whose temperature is to be measured. The pressure is measured, then the corresponding temperature is read off the graph.

FIGURE 16.2b shows the pressure-temperature relationship for three different gases. Notice two important things about this graph.

1. There is a *linear* relationship between temperature and pressure.
2. All gases extrapolate to *zero pressure* at the same temperature: $T_0 = -273°C$. No gas actually gets that cold without condensing, although helium comes very close, but it is surprising that you get the same zero-pressure temperature for any gas and any starting pressure.

The pressure in a gas is due to collisions of the molecules with each other and the walls of the container. A pressure of zero would mean that all motion, and thus all collisions, had ceased. If there were no atomic motion, the system's thermal energy would be zero. The temperature at which all motion would cease, and at which $E_{th} = 0$, is called **absolute zero.** Because temperature is related to thermal energy, absolute zero is the lowest temperature that has physical meaning. We see from the

FIGURE 16.2 The pressure in a constant-volume gas thermometer extrapolates to zero at $T_0 = -273°C$. This is the basis for the concept of absolute zero.

(a)

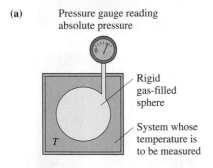

(b)

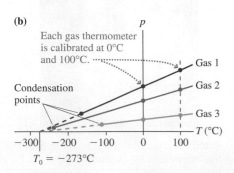

gas-thermometer data that $T_0 = -273°C$. We'll give a somewhat more precise definition of absolute zero in the next section.

It is useful to have a temperature scale with the zero point at absolute zero. Such a temperature scale is called an **absolute temperature scale.** Any system whose temperature is measured on an absolute scale will have $T > 0$. The absolute temperature scale having the same unit size as the Celsius scale is called the *Kelvin scale.* It is the SI scale of temperature. The units of the Kelvin scale are *kelvins,* abbreviated as K. The conversion between the Celsius scale and the Kelvin scale is

$$T_K = T_C + 273 \qquad (16.7)$$

NOTE ▶ The units are simply "kelvins," *not* "degrees Kelvin." ◀

On the Kelvin scale, absolute zero is 0 K, the freezing point of water is 273 K, and the boiling point of water is 373 K. While most practical macroscopic devices utilize temperatures in the range ≈100 K to ≈1000 K, it is worth noting that scientists study the properties of matter from temperatures as low as $\approx10^{-9}$ K (1 nK) on the one extreme to as high as $\approx10^7$ K on the other!

STOP TO THINK 16.3 The temperature of a glass of water increases from 20°C to 30°C. What is ΔT?

a. 10 K   b. 283 K   c. 293 K   d. 303 K

16.4 Phase Changes

The temperature inside the freezer compartment of a refrigerator is typically about −20°C. Suppose you were to remove a few ice cubes from the freezer, place them in a sealed container with a thermometer, then heat them, as **FIGURE 16.3a** shows. We'll assume that the heating is done so slowly that the inside of the container always has a single, well-defined temperature.

FIGURE 16.3b shows the temperature as a function of time. After steadily rising from the initial −20°C, the temperature remains fixed at 0°C for an extended period of time. This is the interval of time during which the ice melts. As it's melting, the ice temperature is 0°C and the liquid water temperature is 0°C. Even though the system is being heated, the liquid water temperature doesn't begin to rise until all the ice has melted. If you were to turn off the flame at any point, the system would remain a mixture of ice and liquid water at 0°C.

NOTE ▶ In everyday language, the three phases of water are called *ice, water,* and *steam.* That is, the term *water* implies the liquid phase. Scientifically, these are the solid, liquid, and gas phases of the compound called *water.* To be clear, we'll use the term *water* in the scientific sense of a collection of H_2O molecules. We'll say either *liquid* or *liquid water* to denote the liquid phase. ◀

The thermal energy of a solid is the kinetic energy of the vibrating atoms plus the potential energy of the stretched and compressed molecular bonds. Melting occurs when the thermal energy gets so large that molecular bonds begin to break, allowing the atoms to move around. The temperature at which a solid becomes a liquid or, if the thermal energy is reduced, a liquid becomes a solid is called the **melting point** or the **freezing point.** Melting and freezing are *phase changes.*

A system at the melting point is in **phase equilibrium,** meaning that any amount of solid can coexist with any amount of liquid. Raise the temperature ever so slightly and the entire system becomes liquid. Lower it slightly and it all becomes solid. But exactly at the melting point the system has no tendency to move one way or the other.

FIGURE 16.3 The temperature as a function of time as water is transformed from solid to liquid to gas.

(a)

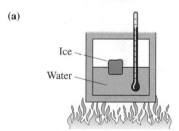

(b)

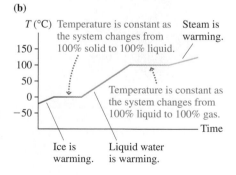

That is why the temperature remains constant at the melting point until the phase change is complete.

You can see the same thing happening in Figure 16.3b at 100°C, the boiling point. This is a phase equilibrium between the liquid phase and the gas phase, and any amount of liquid can coexist with any amount of gas at this temperature. Above this temperature, the thermal energy is too large for bonds to be established between molecules, so the system is a gas. If the thermal energy is reduced, the molecules begin to bond with each other and stick together. In other words, the gas condenses into a liquid. The temperature at which a gas becomes a liquid or, if the thermal energy is increased, a liquid becomes a gas is called the **condensation point** or the **boiling point.**

NOTE ▶ Liquid water becomes solid ice at 0°C, but that doesn't mean the temperature of ice is always 0°C. Ice reaches the temperature of its surroundings. If the air temperature in a freezer is −20°C, then the ice temperature is −20°C. Likewise, steam can be heated to temperatures above 100°C. That doesn't happen when you boil water on the stove because the steam escapes, but steam can be heated far above 100°C in a sealed container. ◀

A **phase diagram** is used to show how the phases and phase changes of a substance vary with both temperature and pressure. **FIGURE 16.4** shows the phase diagrams for water and carbon dioxide. You can see that each diagram is divided into three regions corresponding to the solid, liquid, and gas phases. The boundary lines separating the regions indicate the phase transitions. The system is in phase equilibrium at a pressure-temperature point that falls on one of these lines.

Phase diagrams contain a great deal of information. Notice on the water phase diagram that the dotted line at $p = 1$ atm crosses the solid-liquid boundary at 0°C and the liquid-gas boundary at 100°C. These well-known melting and boiling point temperatures of water apply only at standard atmospheric pressure. You can see that in Denver, where $p_{atmos} < 1$ atm, water melts at slightly above 0°C and boils at a temperature below 100°C. A *pressure cooker* works by allowing the pressure inside to exceed 1 atm. This raises the boiling point, so foods that are in boiling water are at a temperature $> 100°C$ and cook faster.

In general, crossing the solid-liquid boundary corresponds to melting or freezing while crossing the liquid-gas boundary corresponds to boiling or condensing. But there's another possibility—crossing the solid-gas boundary. The phase change in which a solid becomes a gas is called **sublimation.** It is not an everyday experience with water because water sublimates only at pressures far below atmospheric pressure. (It does happen in a vacuum chamber, which is how food products are *freeze dried.*) But you are familiar with the sublimation of dry ice. Dry ice is solid carbon dioxide. You can see on the carbon dioxide phase diagram that the dotted line at $p = 1$ atm crosses the solid-*gas* boundary, rather than the solid-liquid boundary, at $T = -78°C$. This is the *sublimation temperature* of dry ice.

Liquid carbon dioxide does exist, but only at pressures greater than 5 atm and temperatures greater than −56°C. A CO_2 fire extinguisher contains *liquid* carbon dioxide under high pressure. (You can hear the liquid slosh if you shake a CO_2 fire extinguisher.)

One important difference between the water and carbon dioxide phase diagrams is the slope of the solid-liquid boundary. For most substances, the solid phase is denser than the liquid phase and the liquid is denser than the gas. Pressurizing the substance compresses it and increases the density. If you start compressing CO_2 gas at room temperature, thus moving upward through the phase diagram along a vertical line, you'll first condense it to a liquid and eventually, if you keep compressing, change it into a solid.

Water is a very unusual substance in that the density of ice is *less* than the density of liquid water. That is why ice floats. If you compress ice, making it denser, you

FIGURE 16.4 Phase diagrams (not to scale) for water and carbon dioxide.

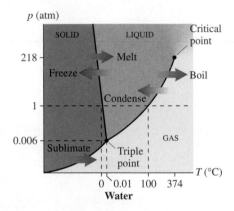

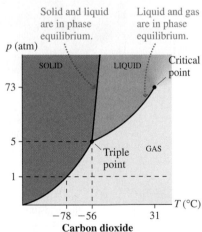

eventually cause a phase transition in which the ice turns to liquid water! Consequently, the solid-liquid boundary for water slopes to the left.

The liquid-gas boundary ends at a point called the **critical point.** Below the critical point, liquid and gas are clearly distinct and there is a phase change if you go from one to the other. But there is no clear distinction between liquid and gas at pressures or temperatures above the critical point. The system is a *fluid,* but it can be varied continuously between high density and low density without a phase change.

The final point of interest on the phase diagram is the **triple point** where the phase boundaries meet. Two phases are in phase equilibrium along the boundaries. The triple point is the *one* value of temperature and pressure for which all three phases can coexist in phase equilibrium. That is, any amounts of solid, liquid, and gas can happily coexist at the triple point. For water, the triple point occurs at $T_3 = 0.01°C$ and $p_3 = 0.006$ atm.

The significance of the triple point of water is its connection to the Kelvin temperature scale. The Celsius scale required two *reference points,* the boiling and melting points of water. We can now see that these are not very satisfactory reference points because their values vary as the pressure changes. In contrast, there's only one temperature at which ice, liquid water, and water vapor will coexist in equilibrium. If you produce this equilibrium in the laboratory, then you *know* the system is at the triple-point temperature.

The triple-point temperature of water is an ideal reference point, hence the Kelvin temperature scale is *defined* to be a linear temperature scale starting from 0 K at absolute zero and passing through 273.16 K at the triple point of water. Because $T_3 = 0.01°C$, absolute zero on the Celsius scale is $T_0 = -273.15°C$.

NOTE ▶ To be consistent with our use of significant figures, $T_0 = -273$ K is the appropriate value to use in calculations *unless* you know other temperatures with an accuracy of better than 1°C. ◀

Food takes longer to cook at high altitudes because the boiling point of water is less than 100°C.

STOP TO THINK 16.4 For which is there a sublimation temperature that is higher than a melting temperature?

a. Water   b. Carbon dioxide   c. Both   d. Neither

16.5 Ideal Gases

Gases are the simplest macroscopic systems. Our goal for the rest of this chapter will be to understand how the macroscopic properties of a gas change as the state of the gas changes.

While we today take it for granted that matter consists of atoms, the evidence for atoms is by no means obvious. The concept of atoms was formulated by two Greek philosophers, Leucippus and Democritus, who flourished about 440–420 BCE. They suggested that all matter consists of small, hard, indivisible, and indestructible particles they called *atoms.*

The atomic model was revived in about 1740 by Daniel Bernoulli, for whom Bernoulli's equation in fluid dynamics is named. He suggested that a gas consists of small, hard atoms moving randomly at fairly high speeds and, on occasion, colliding with each other or the walls of the container. Surprisingly, Bernoulli's ideas were not accepted for nearly a century. The value of his postulates was not recognized until a complete understanding of energy conservation was achieved in the mid-19th century. Numerous scientists then developed Bernoulli's ideas into the kinetic theory of gases that we will study in Chapter 18.

What can macroscopic observations suggest to us about the properties of atoms? One observation, which we noted earlier in the chapter, is that solids and liquids are nearly incompressible. From this we can infer that atoms are fairly "hard" and cannot be pressed together once they come into contact with each other. Atoms also resist being pulled apart. Solids would not be solid if the atoms were not held together by attractive forces. These attractive forces are responsible for the *tensile strength* of solids—how hard you have to pull to break the solid—as well as for the cohesion of liquid droplets. Nonetheless, it is far easier to break a solid or disperse a liquid than to compress it, so these attractive forces must be weak in comparison to the repulsive forces that occur when we push the atoms too close together.

These observations imply that an atom is a small particle that is weakly attracted to other nearby atoms but strongly repelled by them if it gets too close. This is precisely the view of molecular bonds that we developed in Chapter 10. **FIGURE 16.5** shows the potential-energy diagram of two atoms separated by distance r. Recall, from Chapter 11, that the force exerted by one atom on the other is the negative of the slope of this graph. The slope is large and negative for values of r less than the equilibrium value r_{eq}, so the force for $r < r_{eq}$ is large and repulsive. For r just slightly greater than r_{eq}, the modest positive slope indicates a weak attractive force. The slope has become zero by $r \approx 0.4$ nm, hence the attractive force is restricted to atoms within about 0.4 nm of each other. Atoms separated by more than about 0.4 nm do not interact.

Solids and liquids are systems in which the atomic separation is very close to r_{eq}; thus the attractive and repulsive atomic forces are balanced. If you try to press the atoms closer together, the repulsive forces resist. If you try to pull them apart, the attractive forces resist.

A gas, by contrast, is much less dense and the average spacing of atoms is much larger than r_{eq}. Consequently, the atoms are usually *not interacting* with each other at all. Instead, they spend most of their time moving freely through space, only occasionally coming close enough to another atom to interact with it. When two atoms collide, it is the steep "wall" of the potential-energy curve for $r < r_{eq}$ that is important. That wall represents the repulsive electrical force pushing the atoms apart as they collide. The small distance over which the atoms experience a weak attractive force is of essentially no importance because the atoms spend so little time at those distances.

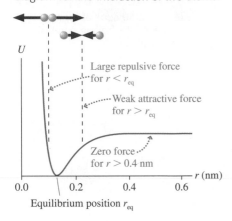

FIGURE 16.5 The potential-energy diagram for the interaction of two atoms.

U

Large repulsive force for $r < r_{eq}$

Weak attractive force for $r > r_{eq}$

Zero force for $r > 0.4$ nm

0.0   0.2   0.4   0.6 r (nm)

Equilibrium position r_{eq}

The Ideal-Gas Model

With these ideas in mind, suppose we were to replace the actual potential-energy curve of Figure 16.5 with the approximate potential-energy curve of **FIGURE 16.6**. This is the potential-energy curve for the interaction of two "hard spheres" that have *no* interaction at all until they come into actual contact, at separation $r_{contact}$, and then bounce.

The *hard-sphere model* of the atom represents what we could call the *ideal atom*. It is Democritus' idea of a small, hard particle. A gas of such ideal atoms is called an **ideal gas.** It is a collection of small, hard, randomly moving atoms that occasionally collide and bounce off each other but otherwise do not interact. The ideal gas is a *model* of a real gas and, as with any other model, a simplified description. Nonetheless, experiments show that the ideal-gas model is quite good for gases if two conditions are met:

1. The density is low (i.e., the atoms occupy a volume much smaller than that of the container), and
2. The temperature is well above the condensation point.

If the density gets too high, or the temperature too low, then the attractive forces between the atoms begin to play an important role and our model, which ignores those attractive forces, fails. These are the forces that are responsible, under the right conditions, for the gas condensing into a liquid.

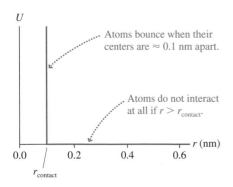

FIGURE 16.6 An idealized hard-sphere model of the interaction potential energy of two atoms.

U

Atoms bounce when their centers are ≈ 0.1 nm apart.

Atoms do not interact at all if $r > r_{contact}$.

0.0   0.2   0.4   0.6 r (nm)

$r_{contact}$

We've been using the term "atoms," but many gases, as you know, consist of molecules rather than atoms. Only helium, neon, argon, and the other inert elements in the far-right column of the periodic table of the elements form monatomic gases. Hydrogen (H_2), nitrogen (N_2), and oxygen (O_2) are diatomic gases. As far as translational motion is concerned, the ideal-gas model does not distinguish between a monatomic gas and a diatomic gas; both are considered as simply small, hard spheres. Hence the terms "atoms" and "molecules" can be used interchangeably to mean the basic constituents of the gas.

The Ideal-Gas Law

Section 16.1 introduced the idea of *state variables,* those parameters that describe the state of a macroscopic system. The state variables for an ideal gas are the volume V of its container, the number of moles n of the gas present in the container, the temperature T of the gas and its container, and the pressure p that the gas exerts on the walls of the container. These four state parameters are not independent of each other. If you change the value of one—by, say, raising the temperature—then one or more of the others will change as well. Each change of the parameters is a *change of state* of the system.

Experiments during the 17th and 18th centuries found a very specific relationship between the four state variables. Suppose you change the state of a gas, by heating it or compressing it or doing something else to it, and measure p, V, n, and T. Repeat this many times, changing the state of the gas each time, until you have a large table of p, V, n, and T values.

Then make a graph on which you plot pV, the product of the pressure and volume, on the vertical axis and nT, the product of the number of moles and temperature (in kelvins), on the horizontal axis. The very surprising result is that for *any* gas, whether it is hydrogen or helium or oxygen or methane, **you get exactly the same graph,** the linear graph shown in **FIGURE 16.7**. In other words, nothing about the graph indicates what gas was used because all gases give the same result.

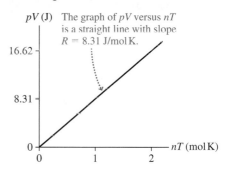

FIGURE 16.7 A graph of pV versus nT for an ideal gas.

NOTE ▶ No real gas could extend to $nT = 0$ because it would condense. But an ideal gas never condenses because the only interactions among the molecules are hard-sphere collisions. ◀

As you can see, there is a very clear proportionality between the quantity pV and the quantity nT. If we designate the slope of the line in this graph as R, then we can write the relationship as

$$pV = R \times (nT)$$

It is customary to write this relationship in a slightly different form, namely

$$pV = nRT \qquad \text{(ideal-gas law)} \qquad (16.8)$$

Equation 16.8 is the **ideal-gas law. The ideal-gas law is a relationship among the four state variables—p, V, n, and T—that characterize a gas in thermal equilibrium.**

The constant R, which is determined experimentally as the slope of the graph in Figure 16.7, is called the **universal gas constant.** Its value, in SI units, is

$$R = 8.31 \text{ J/mol K}$$

The units of R seem puzzling. The denominator mol K is clear because R multiplies nT. But what about the joules? The left side of the ideal-gas law, pV, has units

$$\text{Pa m}^3 = \frac{\text{N}}{\text{m}^2} \text{ m}^3 = \text{N m} = \text{joules}$$

The product pV has units of joules, as shown on the vertical axis in Figure 16.7.

NOTE ▶ You perhaps learned in chemistry to work gas problems using units of atmospheres and liters. To do so, you had a different numerical value of R expressed in those units. In physics, however, we always work gas problems in SI units. Pressures *must* be in Pa, volumes in m^3, and temperatures in K before you compute. Calculations using other units give wildly incorrect answers. ◀

The surprising fact, and one worth commenting upon, is that *all* gases have the *same* graph and the *same* value of R. There is no obvious reason a very simple atomic gas such as helium should have the same slope as a more complex gas such as methane (CH_4). Nonetheless, both turn out to have the same value for R. The ideal-gas law, within its limits of validity, describes *all* gases with a single value of the constant R.

EXAMPLE 16.3 Calculating a gas pressure
100 g of oxygen gas is distilled into an evacuated 600 cm^3 container. What is the gas pressure at a temperature of 150°C?

MODEL The gas can be treated as an ideal gas. Oxygen is a diatomic gas of O_2 molecules.

SOLVE From the ideal-gas law, the pressure is $p = nRT/V$. In Example 16.2 we calculated the number of moles in 100 g of O_2 and found $n = 3.13$ mol. Gas problems typically involve several conversions to get quantities into the proper units, and this example is no exception. The SI units of V and T are m^3 and K, respectively, thus

$$V = (600 \text{ cm}^3)\left(\frac{1 \text{ m}}{100 \text{ cm}}\right)^3 = 6.00 \times 10^{-4} \text{ m}^3$$

$$T = (150 + 273) \text{ K} = 423 \text{ K}$$

With this information, the pressure is

$$p = \frac{nRT}{V} = \frac{(3.13 \text{ mol})(8.31 \text{ J/mol K})(423 \text{ K})}{6.00 \times 10^{-4} \text{ m}^3}$$

$$= 1.83 \times 10^7 \text{ Pa} = 181 \text{ atm}$$

In this text we will consider only gases in sealed containers. The number of moles (and number of molecules) will not change during a problem. In that case,

$$\frac{pV}{T} = nR = \text{constant} \tag{16.9}$$

If the gas is initially in state i, characterized by the state variables p_i, V_i, and T_i, and at some later time in a final state f, the state variables for these two states are related by

$$\frac{p_f V_f}{T_f} = \frac{p_i V_i}{T_i} \quad \text{(ideal gas in a sealed container)} \tag{16.10}$$

This before-and-after relationship between the two states, reminiscent of a conservation law, will be valuable for many problems.

EXAMPLE 16.4 Calculating a gas temperature
A cylinder of gas is at 0°C. A piston compresses the gas to half its original volume and three times its original pressure. What is the final gas temperature?

MODEL Treat the gas as an ideal gas in a sealed container.

SOLVE The before-and-after relationship of Equation 16.10 can be written

$$T_2 = T_1 \frac{p_2}{p_1} \frac{V_2}{V_1}$$

In this problem, the compression of the gas results in $V_2/V_1 = \frac{1}{2}$ and $p_2/p_1 = 3$. The initial temperature is $T_1 = 0°C = 273$ K. With this information,

$$T_2 = 273 \text{ K} \times 3 \times \frac{1}{2} = 409 \text{ K} = 136°C$$

ASSESS We did not need to know actual values of the pressure and volume, just the *ratios* by which they change.

We will often want to refer to the number of molecules N in a gas rather than the number of moles n. This is an easy change to make. Because $n = N/N_A$, the ideal-gas law in terms of N is

$$pV = nRT = \frac{N}{N_A}RT = N\frac{R}{N_A}T \qquad (16.11)$$

R/N_A, the ratio of two known constants, is known as **Boltzmann's constant** k_B:

$$k_B = \frac{R}{N_A} = 1.38 \times 10^{-23} \text{ J/K}$$

The subscript B distinguishes Boltzmann's constant from a spring constant or other uses of the symbol k.

Ludwig Boltzmann was an Austrian physicist who did some of the pioneering work in statistical physics during the mid-19th century. Boltzmann's constant k_B can be thought of as the "gas constant per molecule," whereas R is the "gas constant per mole." With this definition, the ideal-gas law in terms of N is

$$pV = Nk_BT \qquad \text{(ideal-gas law)} \qquad (16.12)$$

Equations 16.8 and 16.12 are both the ideal-gas law, just expressed in terms of different state variables.

Recall that the number density (molecules per m³) was defined as N/V. A rearrangement of Equation 16.12 gives the number density as

$$\frac{N}{V} = \frac{p}{k_BT} \qquad (16.13)$$

This is a useful consequence of the ideal-gas law, but keep in mind that the pressure *must* be in SI units of pascals and the temperature *must* be in SI units of kelvins.

EXAMPLE 16.5 The distance between molecules
"Standard temperature and pressure," abbreviated **STP**, are $T = 0°C$ and $p = 1$ atm. Estimate the average distance between gas molecules at STP.

SOLVE Imagine freezing all the molecules in place at some instant of time. After doing so, place an imaginary cube around each molecule to separate it from all its neighbors. This divides the total volume V of the gas into N small cubes of volume v_i such that the sum of all these small volumes v_i equals the full volume V. Although each of these volumes is somewhat different, we can define an *average* little volume:

$$v_{avg} = \frac{V}{N} = \frac{1}{N/V}$$

That is, the average volume per molecule (m³ per atom) is the inverse of the number of molecules per m³. Note that this is not the volume of the molecule itself, which is much smaller, but the average surrounding volume of space that each molecule can claim as its own, separating it from the other molecules. If we now use Equation 16.13, the number density is

$$\frac{N}{V} = \frac{p}{k_BT} = \frac{1.01 \times 10^5 \text{ Pa}}{(1.38 \times 10^{-23} \text{ J/K})(273 \text{ K})}$$

$$= 2.69 \times 10^{25} \text{ molecules/m}^3$$

where we have used the definition of STP in SI units. Thus the average volume per molecule is

$$v_{avg} = \frac{1}{N/V} = 3.72 \times 10^{-26} \text{ m}^3$$

The volume of a cube is $V = l^3$, where l is the length of each edge. Hence the average length of one of our little cubes is

$$l = (v_{avg})^{1/3} = 3.34 \times 10^{-9} \text{ m} = 3.34 \text{ nm}$$

Because each molecule sits at the center of a cube, the average distance between two molecules is the distance between opposite corners of the cube. As **FIGURE 16.8** shows, this distance is

$$\text{average distance} = \sqrt{l^2 + l^2 + l^2} = \sqrt{3}l = 5.7 \text{ nm}$$

The average distance between molecules in a gas at STP is ≈ 5.7 nm.

FIGURE 16.8 The distance between two molecules.

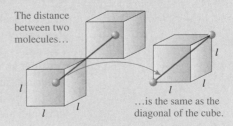

The distance between two molecules… …is the same as the diagonal of the cube.

The results of this example are important. One of the basic assumptions of the ideal-gas model is that the atoms are "far apart" in comparison to the distance over which atoms exert attractive forces on each other. That distance, as was seen in Figure 16.5, is about 0.4 nm. A gas at STP has an average distance between atoms roughly 14 times the interaction distance. We can safely conclude that the ideal-gas model works well for gases under "typical" circumstances.

STOP TO THINK 16.5   You have two containers of equal volume. One is full of helium gas. The other holds an equal mass of nitrogen gas. Both gases have the same pressure. How does the temperature of the helium compare to the temperature of the nitrogen?

a. $T_{\text{helium}} > T_{\text{nitrogen}}$ b. $T_{\text{helium}} = T_{\text{nitrogen}}$ c. $T_{\text{helium}} < T_{\text{nitrogen}}$

16.6 Ideal-Gas Processes

FIGURE 16.9 The state of the gas and ideal-gas processes can be shown on a pV diagram.

(a)   Each state of an ideal gas is represented as a point on a pV diagram.

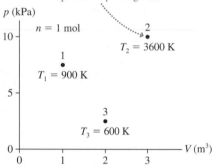

(b)   A process that changes the gas from one state to another is represented by a trajectory on a pV diagram.

(c)   This trajectory represents a different process that takes the gas from state 1 to state 3.

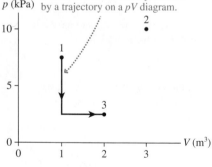

The ideal-gas law is the connection between the state variables pressure, temperature, and volume. If the state variables change, as they would from heating or compressing the gas, the state of the gas changes. An **ideal-gas process** is the means by which the gas changes from one state to another.

NOTE ▶ Even in a sealed container, the ideal-gas law is a relationship among *three* variables. In general, *all three change* during an ideal-gas process. As a result, thinking about cause and effect can be rather tricky. Don't make the mistake of thinking that one variable is constant unless you're sure, beyond a doubt, that it is. ◀

The pV Diagram

It will be very useful to represent ideal-gas processes on a graph called a pV **diagram.** This is nothing more than a graph of pressure versus volume. The important idea behind the pV diagram is that *each point* on the graph represents a single, unique state of the gas. That seems surprising at first, because a point on the graph only directly specifies the values of p and V. But knowing p and V, and assuming that n is known for a sealed container, we can find the temperature by using the ideal-gas law. Thus each point actually represents a triplet of values (p, V, T) specifying the state of the gas.

For example, **FIGURE 16.9a** is a pV diagram showing three states of a system consisting of 1 mol of gas. The values of p and V can be read from the axes for each point, then the temperature at that point determined from the ideal-gas law. An ideal-gas process can be represented as a "trajectory" in the pV diagram. The trajectory shows all the intermediate states through which the gas passes. **FIGURES 16.9b** and **16.9c** show two different processes by which the gas of Figure 16.9a can be changed from state 1 to state 3.

There are infinitely many ways to change the gas from state 1 to state 3. Although the initial and final states are the same for each of them, the particular process by which the gas changes—that is, the particular trajectory—will turn out to have very real consequences. For example, you will soon learn that the work done in compressing gas, a quantity of very practical importance in various devices, depends on the trajectory followed. The pV diagram is an important graphical representation of the process.

Quasi-Static Processes

Strictly speaking, the ideal-gas law applies only to gases in *thermal equilibrium.* We'll give a careful definition of thermal equilibrium later; for now it is sufficient to say that a system is in thermal equilibrium if its state variables are constant and not changing. Consider an ideal-gas process that changes a gas from state 1 to state 2. The initial and final states are states of thermal equilibrium, with steady values of p, V, and T. But the process, by definition, causes some of these state variables to change. The gas is *not* in thermal equilibrium while the process of changing from state 1 to state 2 is under way.

To use the ideal-gas law throughout, we will assume that the process occurs *so slowly* that the system is never far from equilibrium. In other words, the values of p, V, and T at any point in the process are essentially the same as the equilibrium values they would assume if we stopped the process at that point. A process that is essentially in thermal equilibrium at all times is called a **quasi-static process.** It is an idealization, like a frictionless surface, but one that is a very good approximation in many real situations.

An important characteristic of a quasi-static process is that the trajectory through the pV diagram can be *reversed*. If you quasi-statically expand a gas by slowly pulling a piston out, as shown in **FIGURE 16.10a**, you can reverse the process by slowing pushing the piston in. The gas retraces its pV trajectory until it has returned to its initial state. Contrast this with what happens when the membrane bursts in **FIGURE 16.10b**. That is a sudden process, not at all quasi-static. The *irreversible* process of Figure 16.10b cannot be represented on a pV diagram.

The critical question is: How slow must a process be to qualify as quasi-static? That turns out to be a difficult question to answer. This textbook will always assume that processes are quasi-static. It turns out to be a reasonable assumption for the types of examples and homework problems we will look at. Irreversible processes will be left to more advanced courses.

Constant-Volume Process

Many important gas processes take place in a container of constant, unchanging volume. A constant-volume process is called an **isochoric process,** where *iso* is a prefix meaning "constant" or "equal" while *choric* is from a Greek root meaning "volume." An isochoric process is one for which

$$V_f = V_i \qquad (16.14)$$

For example, suppose that you have a gas in the closed, rigid container shown in **FIGURE 16.11a**. Warming the gas with a Bunsen burner will raise its pressure without changing its volume. This process is shown as the vertical line $1 \rightarrow 2$ on the pV diagram of **FIGURE 16.11b**. A constant-volume cooling, by placing the container on a block of ice, would lower the pressure and be represented as the vertical line from 2 to 1. **Any isochoric process appears on a pV diagram as a vertical line.**

FIGURE 16.10 The slow motion of the piston is a quasi-static process. The bursting of the membrane is not.

(a)

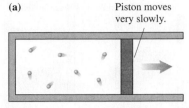

Piston moves very slowly.

Quasi-static process

(b)

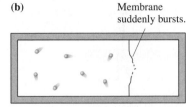

Membrane suddenly bursts.

Irreversible process

FIGURE 16.11 A constant-volume (isochoric) process.

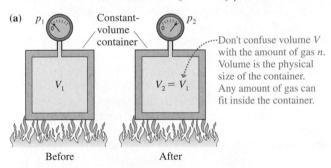

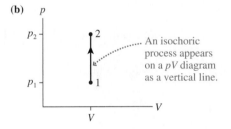

EXAMPLE 16.6 A constant-volume gas thermometer
A constant-volume gas thermometer is placed in contact with a reference cell containing water at the triple point. After reaching equilibrium, the gas pressure is recorded as 55.78 kPa. The thermometer is then placed in contact with a sample of unknown temperature. After the thermometer reaches a new equilibrium, the gas pressure is 65.12 kPa. What is the temperature of this sample?

MODEL The thermometer's volume doesn't change, so this is an isochoric process.

SOLVE The temperature at the triple point of water is $T_1 = 0.01°C = 273.16$ K. The ideal-gas law for a closed system

is $p_2V_2/T_2 = p_1V_1/T_1$. The volume doesn't change, so $V_2/V_1 = 1$. Thus

$$T_2 = T_1 \frac{V_2}{V_1} \frac{p_2}{p_1} = T_1 \frac{p_2}{p_1} = (273.16 \text{ K}) \frac{65.12 \text{ kPa}}{55.78 \text{ kPa}}$$

$$= 318.90 \text{ K} = 45.75°C$$

The temperature *must* be in kelvins to do this calculation, although it is common to convert the final answer to °C. The fact that the pressures were given to four significant figures justified using $T_K = T_C + 273.15$ rather than the usual $T_C + 273$.

ASSESS $T_2 > T_1$, which we expected from the increase in pressure.

FIGURE 16.12 A constant-pressure (isobaric) process.

(a) The piston's mass maintains a constant pressure in the cylinder.

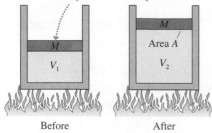

Before   After

(b)

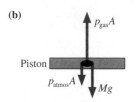

Piston

(c) An isobaric process appears on a *pV* diagram as a horizontal line.

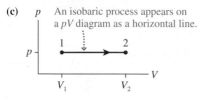

Constant-Pressure Process

Other gas processes take place at a constant, unchanging pressure. A constant-pressure process is called an **isobaric process,** where *baric* is from the same root as "barometer" and means "pressure." An isobaric process is one for which

$$p_f = p_i \qquad (16.15)$$

FIGURE 16.12a shows one method of changing the state of a gas while keeping the pressure constant. A cylinder of gas has a tight-fitting piston of mass M that can slide up and down but seals the container so that no atoms enter or escape. As the free-body diagram of **FIGURE 16.12b** shows, the piston and the air press down with force $p_{atmos}A + Mg$ while the gas inside pushes up with force $p_{gas}A$. In equilibrium, the gas pressure inside the cylinder is

$$p = p_{atmos} + \frac{Mg}{A} \qquad (16.16)$$

In other words, the gas pressure is determined by the requirement that the gas must support both the mass of the piston and the air pressing inward. **This pressure is independent of the temperature of the gas or the height of the piston, so it stays constant as long as M is unchanged.**

If the cylinder is warmed, the gas will expand and push the piston up. But the pressure, determined by mass M, will not change. This process is shown on the *pV* diagram of **FIGURE 16.12c** as the horizontal line $1 \rightarrow 2$. We call this an *isobaric expansion.* An *isobaric compression* occurs if the gas is cooled, lowering the piston. **Any isobaric process appears on a *pV* diagram as a horizontal line.**

EXAMPLE 16.7 Comparing pressure

The two cylinders in **FIGURE 16.13** contain ideal gases at 20°C. Each cylinder is sealed by a frictionless piston of mass M.

a. How does the pressure of gas 2 compare to that of gas 1? Is it larger, smaller, or the same?

b. Suppose gas 2 is warmed to 80°C. Describe what happens to the pressure and volume.

FIGURE 16.13 Compare the pressures of the two gases.

 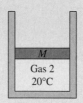

MODEL Treat the gases as ideal gases.

SOLVE a. The pressure in the gas is determined by the requirement that the piston be in mechanical equilibrium. The pressure of the gas inside pushes up on the piston; the air pressure and the weight of the piston press down. The gas pressure $p = p_{atmos} + Mg/A$ depends on the mass of the piston, but not at all on how high the piston is or what type of gas is inside the cylinder. Thus both pressures are the same.

b. Neither does the pressure depend on temperature. Warming the gas increases the temperature, but the pressure—determined by the mass and area of the piston—is unchanged. Because $pV/T = $ constant, and p is constant, it must be true that $V/T = $ constant. As T increases, the volume V also must increase to keep V/T unchanged. In other words, increasing the gas temperature causes the volume to expand—the piston goes up—but with no change in pressure. This is an isobaric process.

EXAMPLE 16.8 A constant-pressure compression

A gas occupying 50.0 cm³ at 50°C is cooled at constant pressure until the temperature is 10°C. What is its final volume?

MODEL The pressure of the gas doesn't change, so this is an isobaric process.

SOLVE By definition, $p_1/p_2 = 1$ for an isobaric process. Using the ideal-gas law for constant n, we have

$$V_2 = V_1 \frac{p_1}{p_2} \frac{T_2}{T_1} = V_1 \frac{T_2}{T_1}$$

Temperatures *must* be in kelvins to use the ideal-gas law. Thus

$$V_2 = (50.0 \text{ cm}^3) \frac{(10 + 273) \text{ K}}{(50 + 273) \text{ K}} = 43.8 \text{ cm}^3$$

ASSESS As long as we use *ratios,* we do not need to convert volumes or pressures to SI units. That is because the conversion is a multiplicative factor that cancels. But the conversion of temperature is an *additive* factor that does *not* cancel. That is why you must always convert temperatures to kelvins in ideal-gas calculations.

Constant-Temperature Process

The last process we wish to look at for now is one that takes place at a constant temperature. A constant-temperature process is called an **isothermal process.** An isothermal process is one for which $T_f = T_i$. Because $pV = nRT$, a constant-temperature process in a closed system (constant n) is one for which the product pV doesn't change. Thus

$$p_f V_f = p_i V_i \qquad (16.17)$$

in an isothermal process.

One possible isothermal process is illustrated in **FIGURE 16.14a,** where a piston is being pushed down to compress a gas. If the piston is pushed *slowly,* then heat energy transfer through the walls of the cylinder will keep the gas at the same temperature as the surrounding liquid. This would be an *isothermal compression.* The reverse process, with the piston slowly pulled out, would be an *isothermal expansion.*

Representing an isothermal process on the pV diagram is a little more complicated than the two previous processes because both p and V change. As long as T remains fixed, we have the relationship

$$p = \frac{nRT}{V} = \frac{\text{constant}}{V} \qquad (16.18)$$

The inverse relationship between p and V causes the graph of an isothermal process to be a *hyperbola*. As one state variable goes up, the other goes down.

The process shown as $1 \rightarrow 2$ in **FIGURE 16.14b** represents the *isothermal compression* shown in Figure 16.14a. An *isothermal expansion* would move in the opposite direction along the hyperbola.

The location of the hyperbola depends on the value of T. A lower-temperature process is represented by a hyperbola closer to the origin than a higher-temperature process. **FIGURE 16.14c** shows four hyperbolas representing the temperatures T_1 to T_4 where $T_4 > T_3 > T_2 > T_1$. These are called **isotherms.** A gas undergoing an isothermal process will move along the isotherm of the appropriate temperature.

Activ Physics 8.4

FIGURE 16.14 A constant-temperature (isothermal) process.

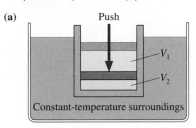
(a)

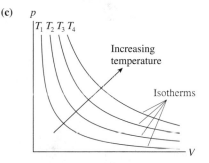
(b)

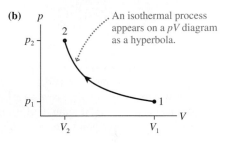
(c)

EXAMPLE 16.9 A constant-temperature compression

A gas cylinder with a tight-fitting, moveable piston contains 200 cm³ of air at 1.0 atm. It floats on the surface of a swimming pool of 15°C water. The cylinder is then slowly pulled underwater to a depth of 3.0 m. What is the volume of gas at this depth?

MODEL The gas's temperature doesn't change, so this is an isothermal compression.

SOLVE At the surface, the pressure inside the cylinder must exactly equal the outside air pressure of 1.0 atm. If the pressures were not equal, a net force would push the piston in or pull it out until the pressures balanced and equilibrium was achieved. As the cylinder is pulled underwater, the increasing water pressure pushes the piston in and compresses the gas. Equilibrium at depth d requires that the gas pressure inside the cylinder equal the water pressure $p_{water} = p_0 + \rho g d$, where $p_0 = 1.0$ atm is the pressure at the surface. As long as the cylinder moves slowly, the gas will stay at the same temperature as the surrounding water. The value of T is not important; all we need to know is that the compression is isothermal. In that case, because $T_2/T_1 = 1$,

$$V_2 = V_1 \frac{T_2}{T_1} \frac{p_1}{p_2} = V_1 \frac{p_1}{p_2} = V_1 \frac{p_0}{p_0 + \rho g d}$$

The initial pressure p_0 must be in SI units: $p_0 = 1.0$ atm $= 1.013 \times 10^5$ Pa. Then a straightforward computation gives $V_2 = 155$ cm³.

ASSESS V_2 is less than V_1. This is expected because the gas is being compressed.

STOP TO THINK 16.6   Two cylinders contain the same number of moles of the same ideal gas. Each cylinder is sealed by a frictionless piston. To have the same pressure in both cylinders, which piston would you use in cylinder 2?

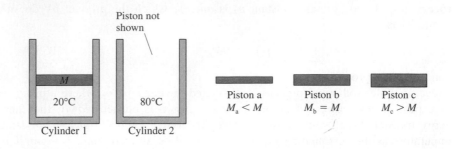

EXAMPLE 16.10 A multistep process

A gas at 2.0 atm pressure and a temperature of 200°C is first expanded isothermally until its volume has doubled. It then undergoes an isobaric compression until it returns to its original volume. First show this process on a pV diagram. Then find the final temperature and pressure.

MODEL Many practical ideal-gas processes consist of several basic steps performed in series. In this case, the final state of the isothermal expansion is the initial state for an isobaric compression.

VISUALIZE FIGURE 16.15 shows the process. The gas starts in state 1 at pressure $p_1 = 2.0$ atm and volume V_1. As the gas expands isothermally, it moves downward along an isotherm until it reaches volume $V_2 = 2V_1$. The pressure decreases during this process to a lower value, p_2. The gas is then compressed at constant pressure p_2 until its final volume V_3 equals its original volume V_1. State 3 is on an isotherm closer to the origin, so we expect to find $T_3 < T_1$.

SOLVE $T_2/T_1 = 1$ during the isothermal expansion and $V_2 = 2V_1$, so the pressure at point 2 is

$$p_2 = p_1 \frac{T_2}{T_1} \frac{V_1}{V_2} = p_1 \frac{V_1}{2V_1} = \frac{1}{2} p_1 = 1.0 \text{ atm}$$

FIGURE 16.15 A pV diagram for the process of Example 16.10.

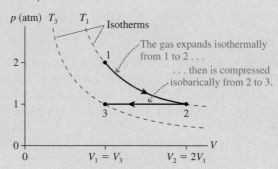

We have $p_3/p_2 = 1$ during the isobaric compression and $V_3 = V_1 = \frac{1}{2} V_2$, so

$$T_3 = T_2 \frac{p_3}{p_2} \frac{V_3}{V_2} = T_2 \frac{\frac{1}{2} V_2}{V_2} = \frac{1}{2} T_2 = 236.5 \text{ K} = -36.5°\text{C}$$

where we converted T_2 to 473 K before doing calculations and then converted T_3 back to °C. The final state, with $T_3 = -36.5°$C and $p_3 = 1.0$ atm, is one in which both the pressure and the absolute temperature are half their original values.

STOP TO THINK 16.7   What is the ratio T_f/T_i for this process?

a. $\frac{1}{4}$

b. $\frac{1}{2}$

c. 1 (no change)

d. 2

e. 4

f. There's not enough information to tell.

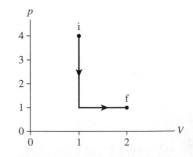

SUMMARY

The goal of Chapter 16 has been to learn the characteristics of macroscopic systems.

General Principles

Three Phases of Matter

Solid   Rigid, definite shape. Nearly incompressible.

Liquid   Molecules loosely held together by molecular bonds, but able to move around. Nearly incompressible.

Gas   Molecules move freely through space. Compressible.

The different phases exist for different conditions of temperature T and pressure p. The boundaries separating the regions of a **phase diagram** are lines of phase equilibrium. Any amounts of the two phases can coexist in equilibrium. The **triple point** is the one value of temperature and pressure at which all three phases can coexist in equilibrium.

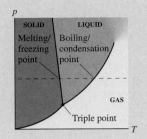

Important Concepts

Ideal-Gas Model

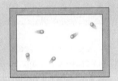

- Atoms and molecules are small, hard spheres that travel freely through space except for occasional collisions with each other or the walls.

- The model is valid when the density is low and the temperature well above the condensation point.

Ideal-Gas Law

The **state variables** of an ideal gas are related by the ideal-gas law

$$pV = nRT \quad \text{or} \quad pV = Nk_B T$$

where $R = 8.31$ J/mol K is the universal gas constant and $k_B = 1.38 \times 10^{-23}$ J/K is Boltzmann's constant.

p, V, and T *must* be in SI units of Pa, m³, and K. For a gas in a sealed container, with constant n:

$$\frac{p_2 V_2}{T_2} = \frac{p_1 V_1}{T_1}$$

Counting atoms and moles

A macroscopic sample of matter consists of N atoms (or molecules), each of mass m (the **atomic** or **molecular mass**):

$$N = \frac{M}{m}$$

Alternatively, we can state that the sample consists of n **moles:**

$$n = \frac{N}{N_A} \quad \text{or} \quad \frac{M(\text{in grams})}{M_{mol}}$$

$N_A = 6.02 \times 10^{23}$ mol⁻¹ is **Avogadro's number.**

The numerical value of the molar mass M_{mol}, in g/mol, equals the numerical value of the atomic or molecular mass m in u. The atomic or molecular mass m, in atomic mass units u, is well approximated by the **atomic mass number** A:

$$1 \text{ u} = 1.66 \times 10^{-27} \text{ kg}$$

The **number density** of the sample is $\dfrac{N}{V}$.

Applications

Temperature scales

$$T_F = \frac{9}{5}T_C + 32° \qquad T_K = T_C + 273$$

The Kelvin temperature scale is based on:

- Absolute zero at $T_0 = 0$ K
- The triple point of water at $T_3 = 273.16$ K

Three basic gas processes

1. **Isochoric,** or constant volume
2. **Isobaric,** or constant pressure
3. **Isothermal,** or constant temperature

pV diagram

Terms and Notation

macroscopic system	atomic mass number, A	absolute temperature scale	Boltzmann's constant, k_B
bulk properties	atomic mass	melting point	STP
micro/macro connection	atomic mass unit, u	freezing point	ideal-gas process
phase	molecular mass	phase equilibrium	pV diagram
phase change	mole, n	condensation point	quasi-static process
solid	monatomic gas	boiling point	isochoric process
crystal	diatomic gas	phase diagram	isobaric process
amorphous solid	Avogadro's number, N_A	sublimation	isothermal process
liquid	molar mass, M_{mol}	critical point	isotherm
gas	temperature, T	triple point	
state variable	constant-volume gas	ideal gas	
thermal equilibrium	thermometer	ideal-gas law	
number density, N/V	absolute zero, T_0	universal gas constant, R	

(MP) For homework assigned on MasteringPhysics, go to www.masteringphysics.com

Problem difficulty is labeled as | (straightforward) to ||| (challenging).

Problems labeled ▣ integrate significant material from earlier chapters.

CONCEPTUAL QUESTIONS

1. Rank in order, from highest to lowest, the temperatures $T_1 = 0$ K, $T_2 = 0°$C, and $T_3 = 0°$F.

2. The sample in an experiment is initially at $10°$C. If the sample's temperature is doubled, what is the new temperature in $°$C?

3. a. Is there a highest temperature at which ice can exist? If so, what is it? If not, why not?
   b. Is there a lowest temperature at which water vapor can exist? If so, what is it? If not, why not?

4. An aquanaut lives in an underwater apartment 100 m beneath the surface of the ocean. Compare the freezing and boiling points of water in the aquanaut's apartment to their values at the surface. Are they higher, lower, or the same? Explain.

5. a. A sample of water vapor in an enclosed cylinder has an initial pressure of 500 Pa at an initial temperature of $-0.01°$C. A piston squeezes the sample smaller and smaller, without limit. Describe what happens to the water as the squeezing progresses.
   b. Repeat part a if the initial temperature is $0.03°$C warmer.

6. The cylinder in **FIGURE Q16.6** is divided into two compartments by a frictionless piston that can slide back and forth. Is the pressure on the left side greater than, less than, or equal to the pressure on the right? Explain.

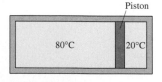

FIGURE Q16.6

7. A gas is in a sealed container. By what factor does the gas pressure change if:
   a. The volume is doubled and the temperature is tripled?
   b. The volume is halved and the temperature is tripled?

8. A gas is in a sealed container. By what factor does the gas temperature change if:
   a. The volume is doubled and the pressure is tripled?
   b. The volume is halved and the pressure is tripled?

9. A gas is in a sealed container. The gas pressure is tripled and the temperature is doubled.
   a. What happens to the number of moles of gas in the container?
   b. What happens to the number density of the gas in the container?

10. A gas undergoes the process shown in **FIGURE Q16.10**. By what factor does the temperature change?

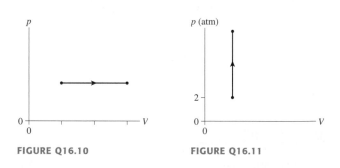

FIGURE Q16.10   **FIGURE Q16.11**

11. The temperature increases from 300 K to 1200 K as a gas undergoes the process shown in **FIGURE Q16.11**. What is the final pressure?

EXERCISES AND PROBLEMS

Exercises

Section 16.1 Solids, Liquids, and Gases

1. | What volume of water has the same mass as 2.0 m^3 of lead?
2. || The nucleus of a uranium atom has a diameter of 1.5×10^{-14} m and a mass of 4.0×10^{-25} kg. What is the density of the nucleus?
3. || What is the diameter of a copper sphere that has the same mass as a 10 cm $\times$ 10 cm $\times$ 10 cm cube of aluminum?
4. || A hollow aluminum sphere with outer diameter 10.0 cm has a mass of 690 g. What is the sphere's inner diameter?

Section 16.2 Atoms and Moles

5. || How many atoms are in a 2.0 cm $\times$ 2.0 cm $\times$ 2.0 cm cube of aluminum?
6. || How many moles are in a 2.0 cm $\times$ 2.0 cm $\times$ 2.0 cm cube of copper?
7. | What is the number density of (a) aluminum and (b) lead?
8. | An element in its solid phase has mass density 1750 kg/m^3 and number density 4.39×10^{28} atoms/m^3. What is the element's atomic mass number?
9. || 1.0 mol of gold is shaped into a sphere. What is the sphere's diameter?
10. || What volume of aluminum has the same number of atoms as 10 cm^3 of mercury?

Section 16.3 Temperature

Section 16.4 Phase Changes

11. | The lowest and highest natural temperatures ever recorded on earth are $-127°$F in Antarctica and 136°F in Libya. What are these temperatures in °C and in K?
12. | At what temperature does the numerical value in °F match the numerical value in °C?
13. || A demented scientist creates a new temperature scale, the "Z scale." He decides to call the boiling point of nitrogen 0°Z and the melting point of iron 1000°Z.
    a. What is the boiling point of water on the Z scale?
    b. Convert 500°Z to degrees Celsius and to kelvins.
14. | What is the temperature in °F and the pressure in Pa at the triple point of (a) water and (b) carbon dioxide?

Section 16.5 Ideal Gases

15. | 3.0 mol of gas at a temperature of $-120°$C fills a 2.0 L container. What is the gas pressure?
16. | A cylinder contains 4.0 g of nitrogen gas. A piston compresses the gas to half its initial volume. Afterward,
    a. Has the mass density of the gas changed? If so, by what factor? If not, why not?
    b. Has the number of moles of gas changed? If so, by what factor? If not, why not?
17. || A rigid container holds 2.0 mol of gas at a pressure of 1.0 atm and a temperature of 30°C.
    a. What is the container's volume?
    b. What is the pressure if the temperature is raised to 130°C?

18. || A gas at 100°C fills volume V_0. If the pressure is held constant, what is the volume if (a) the Celsius temperature is doubled and (b) the Kelvin temperature is doubled?
19. || A 15-cm-diameter compressed-air tank is 50 cm tall. The pressure at 20°C is 150 atm.
    a. How many moles of air are in the tank?
    b. What volume would this air occupy at STP?
20. || A 20-cm-diameter cylinder that is 40 cm long contains 50 g of oxygen gas at 20°C.
    a. How many moles of oxygen are in the cylinder?
    b. How many oxygen molecules are in the cylinder?
    c. What is the number density of the oxygen?
    d. What is the reading of a pressure gauge attached to the tank?
21. || A 10-cm-diameter cylinder of helium gas is 30 cm long and at 20°C. The pressure gauge reads 120 psi.
    a. How many helium atoms are in the cylinder?
    b. What is the mass of the helium?
    c. What is the helium number density?
    d. What is the helium mass density?

Section 16.6 Ideal-Gas Processes

22. | A gas with initial state variables p_1, V_1, and T_1 expands isothermally until $V_2 = 2V_1$. What are (a) T_2 and (b) p_2?
23. | A gas with initial state variables p_1, V_1, and T_1 is cooled in an isochoric process until $p_2 = \frac{1}{3}p_1$. What are (a) V_2 and (b) T_2?
24. || A rigid container holds hydrogen gas at a pressure of 3.0 atm and a temperature of 2°C. What will the pressure be if the temperature is raised to 10°C?
25. || A rigid sphere has a valve that can be opened or closed. The sphere with the valve open is placed in boiling water in a room where the air pressure is 1.0 atm. After a long period of time has elapsed, the valve is closed. What will be the pressure inside the sphere if it is then placed in (a) a mixture of ice and water and (b) an insulated box filled with dry ice?
26. || A 24-cm-diameter vertical cylinder is sealed at the top by a frictionless 20 kg piston. The piston is 84 cm above the bottom when the gas temperature is 303°C.
    a. What is the gas pressure inside the cylinder?
    b. What will the pressure and the height of the piston be if the temperature is lowered to 15°C?
27. || 0.10 mol of argon gas is admitted to an evacuated 50 cm^3 container at 20°C. The gas then undergoes an isochoric heating to a temperature of 300°C.
    a. What is the final pressure of the gas?
    b. Show the process on a pV diagram. Include a proper scale on both axes.
28. | 0.10 mol of argon gas is admitted to an evacuated 50 cm^3 container at 20°C. The gas then undergoes an isobaric heating to a temperature of 300°C.
    a. What is the final volume of the gas?
    b. Show the process on a pV diagram. Include a proper scale on both axes.

29. || 0.10 mol of argon gas is admitted to an evacuated 50 cm³ container at 20°C. The gas then undergoes an isothermal expansion to a volume of 200 cm³.
    a. What is the final pressure of the gas?
    b. Show the process on a pV diagram. Include a proper scale on both axes.

30. | 0.0040 mol of gas undergoes the process shown in **FIGURE EX16.30**.
    a. What type of process is this?
    b. What are the initial and final temperatures in °C?

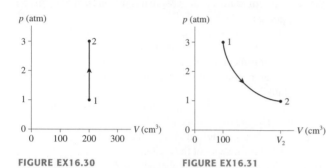

FIGURE EX16.30   **FIGURE EX16.31**

31. || 0.0040 mol of gas undergoes the process shown in **FIGURE EX16.31**.
    a. What type of process is this?
    b. What is the final temperature in °C?
    c. What is the final volume V_2?

32. || A gas with an initial temperature of 900°C undergoes the process shown in **FIGURE EX16.32**.
    a. What type of process is this?
    b. What is the final temperature in °C?
    c. How many moles of gas are there?

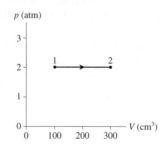

FIGURE EX16.32

Problems

33. || The atomic mass number of copper is $A = 64$. Assume that atoms in solid copper form a cubic crystal lattice. To envision this, imagine that you place atoms at the centers of tiny sugar cubes, then stack the little sugar cubes to form a big cube. If you dissolve the sugar, the atoms left behind are in a cubic crystal lattice. What is the smallest distance between two copper atoms?

34. || An element in its solid phase forms a cubic crystal lattice (see Problem 33) with mass density 7950 kg/m³. The smallest spacing between two adjacent atoms is 0.227 nm. What is the element's atomic mass number?

35. || The molecular mass of water (H_2O) is $A = 18$. How many protons are there in 1.0 L of liquid water?

36. || Estimate the number density of gas molecules in the earth's atmosphere at sea level.

37. || The solar corona is a very hot atmosphere surrounding the visible surface of the sun. X-ray emissions from the corona show that its temperature is about 2×10^6 K. The gas pressure in the corona is about 0.03 Pa. Estimate the number density of particles in the solar corona.

38. || Current vacuum technology can achieve a pressure of 1.0×10^{-10} mm of Hg. At this pressure, and at a temperature of 20°C, how many molecules are in 1 cm³?

39. || The semiconductor industry manufactures integrated circuits in large vacuum chambers where the pressure is 1.0×10^{-10} mm of Hg.
    a. What fraction is this of atmospheric pressure?
    b. At $T = 20$°C, how many molecules are in a cylindrical chamber 40 cm in diameter and 30 cm tall?

40. || A nebula—a region of the galaxy where new stars are forming—contains a very tenuous gas with 100 atoms/cm³. This gas is heated to 7500 K by ultraviolet radiation from nearby stars. What is the gas pressure in atm?

41. || An inflated bicycle inner tube is 2.2 cm in diameter and 200 cm in circumference. A small leak causes the gauge pressure to decrease from 110 psi to 80 psi on a day when the temperature is 20°C. What mass of air is lost? Assume the air is pure nitrogen.

42. || On average, each person in the industrialized world is responsible for the emission of 10,000 kg of carbon dioxide (CO_2) every year. This includes CO_2 that you generate directly, by burning fossil fuels to operate your car or your furnace, as well as CO_2 generated on your behalf by electric generating stations and manufacturing plants. CO_2 is a greenhouse gas that contributes to global warming. If you were to store your yearly CO_2 emissions in a cube at STP, how long would each edge of the cube be?

43. || A gas at 25°C and atmospheric pressure fills a cylinder. The gas is transferred to a new cylinder with three times the volume, after which the pressure is half the original pressure. What is the new temperature of the gas in °C?

44. || On a hot 35°C day, you perspire 1.0 kg of water during your workout.
    a. What volume is occupied by the evaporated water?
    b. By what factor is this larger than the volume occupied by the liquid water?

45. || 10,000 cm³ of 200°C steam at a pressure of 20 atm is cooled until it condenses. What is the volume of the liquid water? Give your answer in cm³.

46. || An electric generating plant boils water to produce high-pressure steam. The steam spins a turbine that is connected to the generator.
    a. How many liters of water must be boiled to fill a 5.0 m³ boiler with 50 atm of steam at 400°C?
    b. The steam has dropped to 2.0 atm pressure at 150°C as it exits the turbine. How much volume does it now occupy?

47. || On a cool morning, when the temperature is 15°C, you measure the pressure in your car tires to be 30 psi. After driving 20 mi on the freeway, the temperature of your tires is 45°C. What pressure will your tire gauge now show?

48. || The air temperature and pressure in a laboratory are 20°C and 1.0 atm. A 1.0 L container is open to the air. The container is then sealed and placed in a bath of boiling water. After reaching thermal equilibrium, the container is opened. How many moles of air escape?

49. | The volume in a constant-*pressure* gas thermometer is directly proportional to the absolute temperature. A constant-pressure thermometer is calibrated by adjusting its volume to 1000 mL while it is in contact with a reference cell at the triple point of water. The volume increases to 1638 mL when the thermometer is placed in contact with a sample. What is the sample's temperature?

50. || The mercury manometer shown in **FIGURE P16.50** is attached to a gas cell. The mercury height h is 120 mm when the cell is placed in an ice-water mixture. The mercury height drops to 30 mm when the device is carried into an industrial freezer. What is the freezer temperature?

    Hint: The right tube of the manometer is much narrower than the left tube. What reasonable assumption can you make about the gas volume?

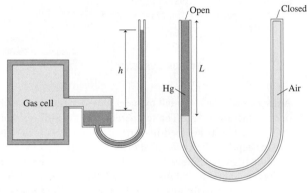

FIGURE P16.50　　　　**FIGURE P16.51**

51. || The U-shaped tube in **FIGURE CP16.51** has a total length of 1.0 m. It is open at one end, closed at the other, and is initially filled with air at 20°C and 1.0 atm pressure. Mercury is poured slowly into the open end without letting any air escape, thus compressing the air. This is continued until the open side of the tube is completely filled with mercury. What is the length L of the column of mercury?

52. || A diver 50 m deep in 10°C fresh water exhales a 1.0-cm-diameter bubble. What is the bubble's diameter just as it reaches the surface of the lake, where the water temperature is 20°C?

    Hint: Assume that the air bubble is always in thermal equilibrium with the surrounding water.

53. || A compressed-air cylinder is known to fail if the pressure exceeds 110 atm. A cylinder that was filled to 25 atm at 20°C is stored in a warehouse. Unfortunately, the warehouse catches fire and the temperature reaches 950°C. Does the cylinder blow?

54. || Reproduce **FIGURE P16.54** on a piece of paper. A gas starts with pressure p_1 and volume V_1. Show on the figure the process in which the gas undergoes an isochoric process that doubles the pressure, then an isobaric process that doubles the volume, followed by an isothermal process that doubles the volume again. Label each of the three processes.

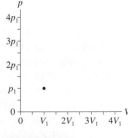

FIGURE P16.54

55. || Reproduce **FIGURE P16.55** on a piece of paper. A gas starts with pressure p_1 and volume V_1. Show on the figure the process in which the gas undergoes an isothermal process during which the volume is halved, then an isochoric process during which the pressure is halved, followed by an isobaric process during which the volume is doubled. Label each of the three processes.

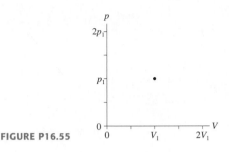

FIGURE P16.55

56. || 8.0 g of helium gas follows the process $1 \rightarrow 2 \rightarrow 3$ shown in **FIGURE P16.56**. Find the values of V_1, V_3, p_2, and T_3.

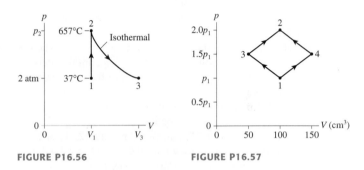

FIGURE P16.56　　　　**FIGURE P16.57**

57. || **FIGURE P16.57** shows two different processes by which 1.0 g of nitrogen gas moves from state 1 to state 2. The temperature of state 1 is 25°C. What are (a) pressure p_1 and (b) temperatures (in °C) T_2, T_3, and T_4?

58. || **FIGURE P16.58** shows two different processes by which 80 mol of gas move from state 1 to state 2. The dashed line is an isotherm.
    a. What is the temperature of the isothermal process?
    b. What maximum temperature is reached along the straight-line process?

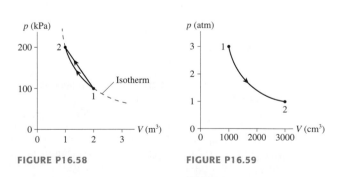

FIGURE P16.58　　　　**FIGURE P16.59**

59. || 0.10 mol of gas undergoes the process $1 \rightarrow 2$ shown in **FIGURE P16.59**.
    a. What are temperatures T_1 and T_2 (in °C)?
    b. What type of process is this?
    c. The gas undergoes an isochoric heating from point 2 until the pressure is restored to the value it had at point 1. What is the final temperature of the gas?

60. ‖ 0.0050 mol of gas undergoes the process $1 \rightarrow 2 \rightarrow 3$ shown in **FIGURE P16.60**. What are (a) temperature T_1, (b) pressure p_2, and (c) volume V_3?

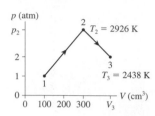

FIGURE P16.60

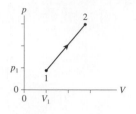

FIGURE P16.61

61. ‖ 4.0 g of oxygen gas, starting at 20°C, follow the process $1 \rightarrow 2$ shown in **FIGURE P16.61**. What is temperature T_2 (in °C)?

62. ‖ 10 g of dry ice (solid CO_2) is placed in a 10,000 cm^3 container, then all the air is quickly pumped out and the container sealed. The container is warmed to 0°C, a temperature at which CO_2 is a gas.
   a. What is the gas pressure? Give your answer in atm. The gas then undergoes an isothermal compression until the pressure is 3.0 atm, immediately followed by an isobaric compression until the volume is 1000 cm^3.
   b. What is the final temperature of the gas (in °C)?
   c. Show the process on a pV diagram.

63. ‖ A container of gas at 2.0 atm pressure and 127°C is compressed at constant temperature until the volume is halved. It is then further compressed at constant pressure until the volume is halved again.
   a. What are the final pressure and temperature of the gas?
   b. Show this process on a pV diagram.

64. ‖ Five grams of nitrogen gas at an initial pressure of 3.0 atm and at 20°C undergo an isobaric expansion until the volume has tripled.
   a. What is the gas volume after the expansion?
   b. What is the gas temperature after the expansion (in °C)? The gas pressure is then decreased at constant volume until the original temperature is reached.
   c. What is the gas pressure after the decrease? Finally, the gas is isothermally compressed until it returns to its initial volume.
   d. What is the final gas pressure?
   e. Show the full three-step process on a pV diagram. Use appropriate scales on both axes.

In Problems 65 through 68 you are given the equation(s) used to solve a problem. For each of these, you are to

   a. Write a realistic problem for which this is the correct equation(s).
   b. Draw a pV diagram.
   c. Finish the solution of the problem.

65. $p_2 = \dfrac{300 \ cm^3}{100 \ cm^3} \times 1 \times 2 \ atm$

66. $(T_2 + 273) \ K = \dfrac{200 \ kPa}{500 \ kPa} \times 1 \times (400 + 273) \ K$

67. $V_2 = \dfrac{(400 + 273) \ K}{(50 + 273) \ K} \times 1 \times 200 \ cm^3$

68. $(2.0 \times 101{,}300 \ Pa)(100 \times 10^{-6} \ m^3) = n(8.31 \ J/mol \ K)T_1$

$$n = \dfrac{0.12 \ g}{20 \ g/mol}$$

$$T_2 = \dfrac{200 \ cm^3}{100 \ cm^3} \times 1 \times T_1$$

Challenge Problems

69. The 50 kg lead piston shown in **FIGURE CP16.69** floats on 0.12 mol of compressed air.
   a. What is the piston height h if the temperature is 30°C?
   b. How far does the piston move if the temperature is increased by 100°C?

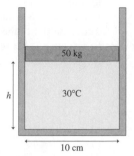

FIGURE CP16.69

70. A diving bell is a 3.0-m-tall cylinder closed at the upper end but open at the lower end. The temperature of the air in the bell is 20°C. The bell is lowered into the ocean until its lower end is 100 m deep. The temperature at that depth is 10°C.
   a. How high does the water rise in the bell after enough time has passed for the air inside to reach thermal equilibrium?
   b. A compressed-air hose from the surface is used to expel all the water from the bell. What minimum air pressure is needed to do this?

71. The 3.0-m-long pipe in **FIGURE CP16.71** is closed at the top end. It is slowly pushed straight down into the water until the top end of the pipe is level with the water's surface. What is the length L of the trapped volume of air?

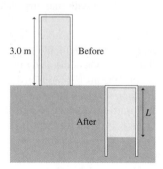

FIGURE CP16.71

72. The cylinder in **FIGURE CP16.72** has a moveable piston attached to a spring. The cylinder's cross-section area is 10 cm^2, it contains 0.0040 mol of gas, and the spring constant is 1500 N/m. At 20°C the spring is neither compressed nor stretched. How far is the spring compressed if the gas temperature is raised to 100°C?

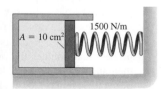

FIGURE CP16.72

73. Containers A and B in **FIGURE CP16.73** hold the same gas. The volume of B is four times the volume of A. The two containers are connected by a thin tube (negligible volume) and a valve that is closed. The gas in A is at 300 K and pressure of 1.0×10^5 Pa. The gas in B is at 400 K and pressure of 5.0×10^5 Pa. Heaters will maintain the temperatures of A and B even after the valve is opened.

    a. After the valve is opened, gas will flow one way or the other until A and B have equal pressure. What is this final pressure?
    b. Is this a reversible or an irreversible process? Explain.

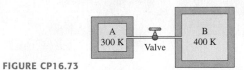

FIGURE CP16.73

74. The closed cylinder of **FIGURE CP16.74** has a tight-fitting but frictionless piston of mass M. The piston is in equilibrium when the left chamber has pressure p_0 and length L_0 while the spring on the right is compressed by ΔL.

    a. What is ΔL in terms of p_0, L_0, A, M, and k?
    b. Suppose the piston is moved a small distance x to the right. Find an expression for the net force $(F_x)_{net}$ on the piston. Assume all motions are slow enough for the gas to remain at the same temperature as its surroundings.
    c. If released, the piston will oscillate around the equilibrium position. Assuming $x \ll L_0$ find an expression for the oscillation period T.
        Hint: Use the binomial approximation.

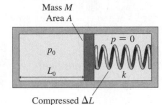

FIGURE CP16.74   Compressed ΔL

STOP TO THINK ANSWERS

Stop to Think 16.1: d. The pressure *decreases* by 20 kPa.

Stop to Think 16.2: a. The number of atoms depends only on the number of moles, not the substance.

Stop to Think 16.3: a. The step size on the Kelvin scale is the same as the step size on the Celsius scale. A *change* of 10°C is a *change* of 10 K.

Stop to Think 16.4: a. On the water phase diagram, you can see that for a pressure just slightly below the triple-point pressure, the solid/gas transition occurs at a higher temperature than does the solid/liquid transition at high pressures. This is not true for carbon dioxide.

Stop to Think 16.5: c. $T = pV/nR$. Pressure and volume are the same, but n differs. The number of moles in mass M is $n = M/M_{mol}$. Helium, with the smaller molar mass, has a larger number of moles and thus a lower temperature.

Stop to Think 16.6: b. The pressure is determined entirely by the weight of the piston pressing down. Changing the temperature changes the volume of the gas, but not its pressure.

Stop to Think 16.7: b. The temperature decreases by a factor of 4 during the isochoric process, where $p_f/p_i = \frac{1}{4}$. The temperature then increases by a factor of 2 during the isobaric expansion, where $V_f/V_i = 2$.

17 Work, Heat, and the First Law of Thermodynamics

This false-color thermal image—an infrared photo—shows where heat energy is escaping from the house.

▶ Looking Ahead

The goals of Chapter 17 are to expand our understanding of energy and to develop the first law of thermodynamics as a general statement of energy conservation. In this chapter you will learn to:

- Understand the energy transfers known as *work* and *heat.*
- Use the first law of thermodynamics.
- Calculate work and heat for ideal-gas processes.
- Use specific heats and heats of transformation in the practical application of calorimetry.
- Understand adiabatic processes.

◀ Looking Back

The material in this chapter continues the development of energy ideas from Chapter 11. Many of the examples depend on the properties of ideal gases. Please review:

- Section 11.4 Work.
- Sections 11.7 and 11.8 Conservation of energy.
- Sections 16.4–16.6 Phase changes and ideal gases.

The industrial revolution was powered by the steam engine. Heat from a wood or coal fire was used to boil water and produce high-pressure steam. The expanding steam pushed a piston that, through a series of gears and levers, turned paddle wheels, ran machinery, or even powered massive locomotives. Humans had used heat for thousands of years for activities ranging from cooking to metallurgy, but the steam engine marked the first time in human history that heat was used to do work.

Our goal in this chapter is to investigate the connection between work and heat in macroscopic systems. Work and heat are *energy transfers* between the system and its environment, so we will be continuing the development of energy concepts that we began in Chapters 10 and 11. In addition, we will want to understand how the state of a system *changes* in response to work and heat. These two ideas, the transfer of energy and the change in the system, are related to each other through the *first law of thermodynamics,* a powerful statement about energy conservation.

17.1 It's All About Energy

A key idea of Chapter 11 was the work-kinetic energy theorem in the form

$$\Delta K = W_c + W_{\text{diss}} + W_{\text{ext}} \qquad (17.1)$$

Equation 17.1 tells us that the kinetic energy of a system of particles is changed when forces do work on the particles by pushing or pulling them through a distance. Here

1. W_c is the work done by conservative forces. This work can be represented as a change in the system's potential energy: $\Delta U = -W_c$.
2. W_{diss} is the work done by friction-like dissipative forces within the system. This work increases the system's thermal energy: $\Delta E_{\text{th}} = -W_{\text{diss}}$.
3. W_{ext} is the work done by external forces that originate in the environment. The push of a piston rod would be an external force.

With these definitions, Equation 17.1 becomes

$$\Delta K + \Delta U + \Delta E_{\text{th}} = W_{\text{ext}} \qquad (17.2)$$

The system's *mechanical energy* was defined as $E_{\text{mech}} = K + U$. **FIGURE 17.1** reminds you that the mechanical energy is associated with the motion of the system as a whole, while E_{th} is associated with the motion of the atoms and molecules within the system. E_{mech} is the *macroscopic* energy of the system as a whole while E_{th} is the *microscopic* energy of the particle-like atoms and spring-like molecular bonds. This led to our final energy statement of Chapter 11:

$$\Delta E_{\text{sys}} = \Delta E_{\text{mech}} + \Delta E_{\text{th}} = W_{\text{ext}} \qquad (17.3)$$

Thus the total energy of an *isolated system,* for which $W_{\text{ext}} = 0$, is constant. This was the essence of the law of conservation of energy as stated in Chapter 11.

The emphasis in Chapters 10 and 11 was on isolated systems. There we were interested in learning how kinetic and potential energy were *transformed* into each other and, where there is friction, into thermal energy. Now we want to focus on how energy is *transferred* between the system and its environment, when W_{ext} is *not* zero.

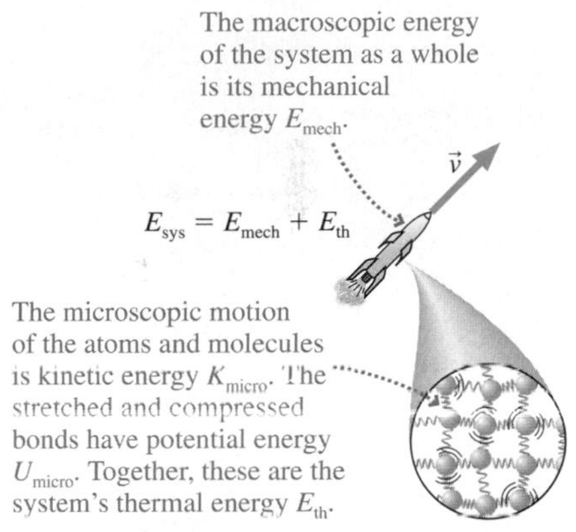

Thermal Energy

Thermal energy, seen in the blow-up of Figure 17.1, is the sum of K_{micro}, the kinetic energy of all the moving atoms and molecules, and U_{micro}, the potential energy stored in the spring-like molecular bonds. That is,

$$E_{\text{th}} = K_{\text{micro}} + U_{\text{micro}} \qquad (17.4)$$

Thermal energy may be hidden from our macroscopic view, but it's quite real. Recall, from Chapter 16, that thermal energy is associated with the system's temperature.

Strictly speaking, the thermal energy due to molecular motion is only one form of energy that can be stored within a system at the microscopic level. For example, a system might have *chemical energy* that can be released via chemical reactions between molecules in the system. Chemical energy is quite important in engineering thermodynamics, where it is needed to characterize combustion processes. *Nuclear energy* is stored in the atomic nuclei and can be released during radioactive decay. All the sources of microscopic energy taken together are called the system's **internal energy:**

$$E_{\text{int}} = E_{\text{th}} + E_{\text{chem}} + E_{\text{nuc}} + \cdots \qquad (17.5)$$

The total energy of the system is then $E_{\text{sys}} = E_{\text{mech}} + E_{\text{int}}$. This textbook will concentrate on simple thermodynamic systems in which the internal energy is entirely thermal: $E_{\text{int}} = E_{\text{th}}$. We'll leave other forms of internal energy to more advanced courses.

FIGURE 17.1 The total energy of a system consists of the macroscopic mechanical energy of the system as a whole plus the microscopic thermal energy of the atoms.

The macroscopic energy of the system as a whole is its mechanical energy E_{mech}.

$$E_{\text{sys}} = E_{\text{mech}} + E_{\text{th}}$$

The microscopic motion of the atoms and molecules is kinetic energy K_{micro}. The stretched and compressed bonds have potential energy U_{micro}. Together, these are the system's thermal energy E_{th}.

FIGURE 17.2 The work done by tension can have very different consequences.

(a) Lift at steady speed

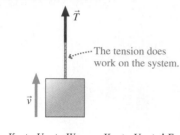

The tension does work on the system.

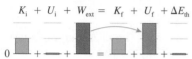

$$K_i + U_i + W_{ext} = K_f + U_f + \Delta E_{th}$$

The energy transferred to the system goes entirely to the system's mechanical energy.

(b) Drag at steady speed

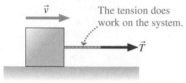

The tension does work on the system.

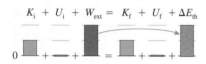

$$K_i + U_i + W_{ext} = K_f + U_f + \Delta E_{th}$$

The energy transferred to the system goes entirely to the system's thermal energy.

Energy Transfer

Doing work on a system can have very different consequences. **FIGURE 17.2a** shows an object being lifted at steady speed by a rope. The rope's tension is an external force doing work W_{ext} on the system. In this case, the energy transferred into the system goes entirely to increasing the system's macroscopic potential energy U_{grav}, part of the mechanical energy. The energy-transfer process $W_{ext} \rightarrow E_{mech}$ is shown graphically in the energy bar chart of Figure 17.2a.

Contrast this with **FIGURE 17.2b**, where the same rope with the same tension now drags the object at steady speed across a rough surface. The tension does the same amount of work, but the mechanical energy does not change. Instead, friction increases the thermal energy of the object + surface system. The energy-transfer process $W_{ext} \rightarrow E_{th}$ is shown in the energy bar chart of Figure 17.2b.

The point of this example is that the energy transferred to a system can go entirely to the system's mechanical energy, entirely to its thermal energy, or (imagine dragging the object up an incline) some combination of the two. The energy isn't lost, but where it ends up depends on the circumstances.

That Can't Be All

You can transfer energy into a system by the mechanical process of doing work on the system. But that can't be all there is to energy transfer. What happens when you place a pan of water on the stove and light the burner? The water temperature increases, so $\Delta E_{th} > 0$. But no work is done ($W_{ext} = 0$) and there is no change in the water's mechanical energy ($\Delta E_{mech} = 0$). This process clearly violates the energy equation $\Delta E_{mech} + \Delta E_{th} = W_{ext}$. What's wrong?

Nothing is wrong. The energy equation is correct as far as it goes, but it is incomplete. Work is energy transferred in a mechanical interaction, but that is not the only way a system can interact with its environment. Energy can also be transferred between the system and the environment if they have a *thermal interaction*. The energy transferred in a thermal interaction is called *heat*.

The symbol for heat is Q. When heat is included, the energy equation becomes

$$\Delta E_{sys} = \Delta E_{mech} + \Delta E_{th} = W + Q \qquad (17.6)$$

Heat and work, now on an equal footing, are both energy transferred between the system and the environment.

> **NOTE** ► We've dropped the subscript "ext" from W. The work that we consider in thermodynamics is *always* the work done by the environment on the system. We won't need to distinguish this work from W_c or W_{diss}, so the subscript is superfluous. ◄

We'll return to Equation 17.6 in Section 17.4 after we look at how work is calculated for ideal-gas processes and at what heat is.

STOP TO THINK 17.1 A gas cylinder and piston are covered with heavy insulation. The piston is pushed into the cylinder, compressing the gas. In this process the gas temperature

a. Increases.
b. Decreases.
c. Doesn't change.
d. There's not sufficient information to tell.

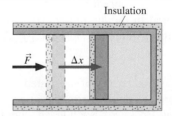

Insulation

17.2 Work in Ideal-Gas Processes

We introduced the idea of *work* in Chapter 11. **Work** is the energy transferred between a system and the environment when a net force acts on the system over a distance. The process itself is a **mechanical interaction,** meaning that the system and the environment interact via macroscopic pushes and pulls. Loosely speaking, we say that the environment (or a particular force from the environment) "does work" on the system. A system is in **mechanical equilibrium** if there is no net force on the system.

FIGURE 17.3 reminds you that work can be either positive or negative. **The sign of the work is *not* just an arbitrary convention, nor does it have anything to do with the choice of coordinate system.** The sign of the work tells us which way energy is being transferred.

FIGURE 17.3 The sign of work.

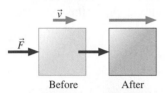 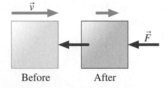

Work is *positive* when the force is in the direction of motion.
■ The force causes the object to speed up.
■ Energy is transferred from the environment to the system.
■ The system's energy increases.

Work is *negative* when the force is opposite to the motion.
■ The force causes the object to slow down.
■ Energy is transferred from the system to the environment.
■ The system's energy decreases.

The pistons in a car engine do work on the air-fuel mixture by compressing it.

In contrast to the mechanical energy or the thermal energy, **work is not a state variable.** That is, work is not a number characterizing the system. Instead, work is the amount of energy that moves between the system and the environment during a mechanical interaction. We can measure the *change* in a state variable, such as a temperature change $\Delta T = T_f - T_i$, but it would make no sense to talk about a "change of work." Consequently, work always appears as W, never as ΔW.

You learned in Chapter 11 how to calculate work. The small amount of work dW done by force $\vec{F}$ as a system moves through the small displacement $d\vec{s}$ is $dW = \vec{F} \cdot d\vec{s}$. If we restrict ourselves to situations where $\vec{F}$ is either parallel or opposite to $d\vec{s}$, then the total work done on the system as it moves from s_i to s_f is

$$W = \int_{s_i}^{s_f} F_s \, ds \tag{17.7}$$

Let's apply this definition to a gas as it expands or is compressed. **FIGURE 17.4a** shows a gas cylinder sealed at one end by a movable piston. Force $\vec{F}_{ext}$, perhaps a force supplied by a piston rod, is equal in magnitude and opposite in direction to $\vec{F}_{gas}$. The gas pressure would blow the piston out of the cylinder if the external force weren't there! Using the coordinate system of Figure 17.4a,

$$(F_{ext})_x = -(F_{gas})_x = -pA \tag{17.8}$$

Suppose the piston moves the small distance dx shown in **FIGURE 17.4b**. As it does so, the external force (i.e., the environment) does work

$$dW = (F_{ext})_x \, dx = -pA \, dx \tag{17.9}$$

If dx is positive (the gas expands), then dW is negative. This is because the external force is opposite the displacement. dW is positive if the gas is slightly compressed (negative dx) because the force and the displacement are in the same direction. This is an important idea.

FIGURE 17.4 The external force does work on the gas as the piston moves.

(a) The gas pushes on the piston with force $\vec{F}_{gas}$.

To keep the piston in place, an external force must be equal and opposite to $\vec{F}_{gas}$.

Pressure p

$\vec{F}_{gas}$ $\vec{F}_{ext}$

0   Piston area A

(b) As the piston moves dx, the external force does work $(F_{ext})_x \, dx$ on the gas.

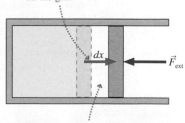

dx $\vec{F}_{ext}$

The volume changes by $dV = A \, dx$ as the piston moves dx.

NOTE ▶ The force $\vec{F}_{gas}$ due to the gas pressure inside the cylinder also does work. Because $\vec{F}_{gas} = -\vec{F}_{ext}$, by Newton's third law, the work done by the gas is simply $W_{gas} = -W_{ext}$. To compress the gas, the environment does positive work and the gas does negative work. As the gas expands, W_{gas} is positive and W_{ext} is negative. But the work that appeared in the work-kinetic energy theorem, and now appears in the laws of thermodynamics, is the work done *on* the system by external forces, not the work done *by* the system. It is W_{ext} that tells us whether energy enters the system or leaves the system—by whether it is positive or negative—and that is why we focus our attention on W_{ext} rather than on W_{gas}. ◀

As the piston moves dx, the volume of the gas changes by $dV = A\,dx$. Consequently, Equation 17.9 can be written in terms of the cylinder's volume as

$$dW = -p\,dV \tag{17.10}$$

If we let the piston move in a slow quasi-static process from initial volume V_i to final volume V_f, the total work done by the environment on the gas is found by integrating Equation 17.10:

$$W = -\int_{V_i}^{V_f} p\,dV \qquad \text{(work done on a gas)} \tag{17.11}$$

Equation 17.11 is a key result of thermodynamics. Although we used a cylinder to derive Equation 17.11, it turns out to be true for a container of any shape.

NOTE ▶ The pressure of a gas usually changes as the gas expands or contracts. Consequently, p is *not* a constant that can be brought outside the integral. You need to know how the pressure changes with volume before you can carry out the integration. ◀

We can give the work done on a gas a nice geometric interpretation. You learned in Chapter 16 how to represent an ideal-gas process as a curve in the pV diagram. Figure 17.5 shows that the work done on a gas is the negative of the area under the pV curve as the volume changes from V_i to V_f. That is

$W =$ the negative of the area under the pV curve between V_i and V_f

FIGURE 17.5a shows a process in which a gas *expands* from V_i to a larger volume V_f. The area under the curve is positive, so the environment does a negative amount of work on an expanding gas. **FIGURE 17.5b** shows a process in which a gas is compressed to a smaller volume. This one is a little trickier because we have to integrate "backward" along the V-axis. You learned in calculus that integrating from a larger limit to a smaller limit gives a negative result, so the area in Figure 17.5b is a negative area. Consequently, as the minus sign in Equation 17.11 indicates, the environment does positive work on a gas to compress it.

FIGURE 17.5 The work done on a gas is the negative of the area under the curve.

(a)   For an *expanding* gas ($V_f > V_i$), the area under the pV curve is positive (integration direction is to the right). Thus the environment does *negative* work on the gas.

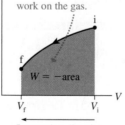

Integration direction

(b)   For a *compressed* gas ($V_f < V_i$), the area is negative because the integration direction is to the left. Thus the environment does *positive* work on the gas.

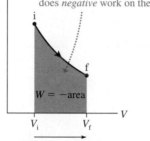

Integration direction

EXAMPLE 17.1 The work done on an expanding gas

How much work is done on the gas in the ideal-gas process of **FIGURE 17.6?**

MODEL The work done on a gas is the negative of the area under the pV curve. The gas is *expanding,* so we expect the work to be negative.

SOLVE The work W is the negative of the area under the curve from $V_i = 500\ \text{cm}^3$ to $V_f = 1500\ \text{cm}^3$. Volumes *must* be converted to SI units of m^3. The area from $500\ \text{cm}^3$ to $1000\ \text{cm}^3$ can be divided into a rectangle (between 0 kPa and 100 kPa) and a triangle (between 100 and 300 kPa). This area is

FIGURE 17.6 The ideal-gas process of Example 17.1.

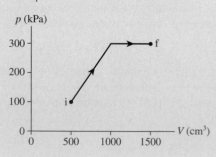

Area$(500 \rightarrow 1000 \text{ cm}^3)$

$$= ((1000 - 500) \times 10^{-6} \text{ m}^3)(100,000 \text{ Pa} - 0 \text{ Pa})$$

$$+ \frac{1}{2}((1000 - 500) \times 10^{-6} \text{ m}^3)$$

$$\times (300,000 \text{ Pa} - 100,000 \text{ Pa})$$

$$= 100 \text{ J}$$

The area from 1000 cm³ to 1500 cm³ is a rectangle:

Area$(1000 \rightarrow 1500 \text{ cm}^3)$

$$= ((1500 - 1000) \times 10^{-6} \text{ m}^3)(300,000 \text{ Pa} - 0 \text{ Pa})$$

$$= 150 \text{ J}$$

The total area under the curve is 250 J, so the work done on the gas as it expands is

$$W = -(\text{area under the } pV \text{ curve}) = -250 \text{ J}$$

ASSESS We noted previously that the product Pa m³ is equivalent to joules. The work is negative, as expected, because the external force pushing on the piston is opposite the direction of the piston's displacement.

Equation 17.11 is the basis for a problem-solving strategy.

PROBLEM-SOLVING STRATEGY 17.1 **Work in ideal-gas processes**

MODEL Assume the gas is ideal and the process is quasi-static.

VISUALIZE Show the process on a pV diagram. Note whether it happens to be one of the basic gas processes: isochoric, isobaric, or isothermal.

SOLVE Calculate the work as the area under the pV curve either geometrically or by carrying out the integration:

Work done on the gas $W = -\displaystyle\int_{V_i}^{V_f} p \, dV = -(\text{area under } pV \text{ curve})$

ASSESS Check your signs.

- $W > 0$ when the gas is compressed. Energy is transferred from the environment to the gas.
- $W < 0$ when the gas expands. Energy is transferred from the gas to the environment.
- No work is done if the volume doesn't change. $W = 0$.

Isochoric Process

The isochoric process in **FIGURE 17.7a** is one in which the volume does not change. Consequently,

$$W = 0 \quad \text{(isochoric process)} \quad (17.12)$$

An isochoric process is the *only* ideal-gas process in which no work is done.

Isobaric Process

FIGURE 17.7b shows an isobaric process in which the volume changes from V_i to V_f. The rectangular area under the curve is $p\Delta V$, so the work done during this process is

$$W = -p\,\Delta V \quad \text{(isobaric process)} \quad (17.13)$$

where $\Delta V = V_f - V_i$. ΔV is positive if the gas expands $(V_f > V_i)$, so W is negative. ΔV is *negative* if the gas is compressed $(V_f < V_i)$, making W positive.

FIGURE 17.7 Calculating the work done during ideal-gas processes.

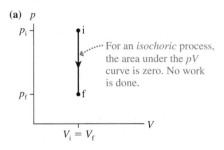

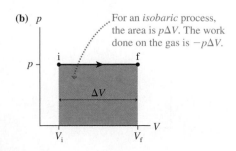

Isothermal Process

FIGURE 17.8 An isothermal process.

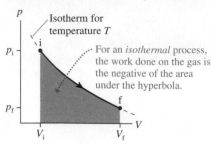

FIGURE 17.8 shows an isothermal process. Here we need to know the pressure as a function of volume before we can integrate Equation 17.11. From the ideal-gas law, $p = nRT/V$. Thus the work on the gas as the volume changes from V_i to V_f is

$$W = -\int_{V_i}^{V_f} p\, dV = -\int_{V_i}^{V_f} \frac{nRT}{V}\, dV = -nRT\int_{V_i}^{V_f} \frac{dV}{V} \tag{17.14}$$

where we could take the T outside the integral because temperature is constant during an isothermal process. This is a straightforward integration, giving

$$W = -nRT\int_{V_i}^{V_f} \frac{dV}{V} = -nRT\,\ln V\Big|_{V_i}^{V_f}$$
$$= -nRT(\ln V_f - \ln V_i) = -nRT\ln\left(\frac{V_f}{V_i}\right) \tag{17.15}$$

Because $nRT = p_iV_i = p_fV_f$ during an isothermal process, the work is:

$$W = -nRT\ln\left(\frac{V_f}{V_i}\right) = -p_iV_i\ln\left(\frac{V_f}{V_i}\right) = -p_fV_f\ln\left(\frac{V_f}{V_i}\right) \tag{17.16}$$
(isothermal process)

8.5 Act|v ONLINE Physics

Which version of Equation 17.16 is easiest to use will depend on the information you're given. The pressure, volume, and temperature *must* be in SI units.

EXAMPLE 17.2 The work of an isothermal compression

A cylinder contains 7.0 g of nitrogen gas. How much work must be done to compress the gas at a constant temperature of 80°C until the volume is halved?

MODEL This is an isothermal ideal-gas process.

SOLVE Nitrogen gas is N_2, with molar mass $M_{mol} = 28$ g/mol, so 7.0 g is 0.25 mol of gas. The temperature is $T = 353$ K. Although we don't know the actual volume, we do know that $V_f = \frac{1}{2}V_i$. The volume ratio is all we need to calculate the work:

$$W = -nRT\ln\left(\frac{V_f}{V_i}\right)$$
$$= -(0.25\text{ mol})(8.31\text{ J/mol K})(353\text{ K})\ln(1/2) = 508\text{ J}$$

ASSESS The work is positive because a force from the environment pushes the piston inward to compress the gas.

FIGURE 17.9 The work done during an ideal-gas process depends on the path.

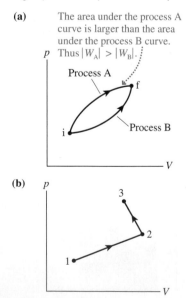

(a) The area under the process A curve is larger than the area under the process B curve. Thus $|W_A| > |W_B|$.

(b)

Work Depends on the Path

FIGURE 17.9a shows two different processes that take a gas from an initial state i to a final state f. Although the initial and final states are the same, the work done during these two processes is *not* the same. **The work done during an ideal-gas process depends on the path followed through the pV diagram.**

You may be thinking that work is supposed to be independent of the path, but that is not the case here. The path we considered in Chapter 11 was the trajectory of a particle from one point to another through space. For an ideal-gas process, the "path" is a sequence of thermodynamic states on a pV diagram. It is a figurative path because we can draw a picture of it on a pV diagram, but it is not a literal path.

The path dependence of work has an important implication for multistep processes such as the one shown in **FIGURE 17.9b**. The total work done on the gas during the process $1 \rightarrow 2 \rightarrow 3$ must be calculated as $W_{1\text{ to }3} = W_{1\text{ to }2} + W_{2\text{ to }3}$. In this case, $W_{1\text{ to }2}$ is negative and $W_{2\text{ to }3}$ is positive. Trying to compute the work in a single step, using

$\Delta V = V_3 - V_1$, would give you the work of a process that goes directly from 1 to 3. The initial and final states are the same, but the work is *not* the same because work depends on the path followed through the pV diagram.

STOP TO THINK 17.2 Two processes take an ideal gas from state 1 to state 3. Compare the work done by process A to the work done by process B.

a. $W_A = W_B = 0$
b. $W_A = W_B$ but neither is zero
c. $W_A > W_B$
d. $W_A < W_B$

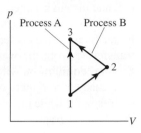

17.3 Heat

Heat is a more elusive concept than work. We use the word "heat" very loosely in the English language, often as synonymous with *hot*. We might say on a very hot day, "This heat is oppressive." If your apartment is cold, you may say, "Turn up the heat." These expressions date to a time long ago when it was thought that heat was a *substance* with fluid-like properties.

One of the first to disagree with the notion of heat as a substance, in the late 1700s, was the American-born Benjamin Thompson. Thompson fled to Europe during the American Revolution, settling in Bavaria and later receiving the title Count Rumford. There, while watching the hot metal chips thrown off during the boring of cannons, he began to think about heat. If heat is a substance, the cannon and borer should eventually run out of heat. But Rumford noted that the heat generation appears to be "inexhaustible," which is not consistent with the idea of heat as a substance. He concluded that heat is not a substance—it is *motion!*

Rumford was beginning to think along the same lines as had Bernoulli. But Rumford's ideas were speculative and qualitative, hardly a scientific theory, and their implications were not immediately grasped by others. Like Bernoulli's, it would be some time before his insight was recognized and validated.

The turning point was the work of British physicist James Joule in the 1840s. Unlike Bernoulli and Count Rumford, Joule carried out careful experiments to learn how it is that systems change their temperature. Using experiments like those shown in FIGURE 17.10, Joule found that you can raise the temperature of a beaker of water by two entirely different means:

1. Heating it with a flame, or
2. Doing work on it with a rapidly spinning paddle wheel.

The final state of the water is *exactly* the same in both cases. This implies that heat and work are essentially equivalent. In other words, heat is not a substance. Instead, heat is *energy*.

Heat and work, which previously had been regarded as two completely different phenomena, were now seen to be simply two different ways of transferring energy to or from a system. Joule's discoveries vindicated the earlier ideas of Bernoulli and Count Rumford, and they opened the door for rapid advancements in the subject of thermodynamics during the second half of the 19th century.

Thermal Interactions

To be specific, **heat** is the energy transferred between a system and the environment as a consequence of a *temperature difference* between them. Unlike a mechanical interaction in which work is done, heat requires no macroscopic motion of the system. Instead

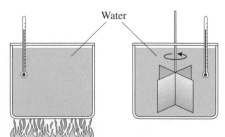

Heat is the energy transferred in a thermal interaction.

FIGURE 17.10 Joule's experiments to show the equivalence of heat and work.

The flame heats the water. The temperature increases.

The spinning paddle does work on the water. The temperature increases.

FIGURE 17.11 The sign of heat.

(a) Positive heat

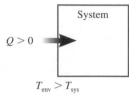

$Q > 0$

$T_{env} > T_{sys}$

(b) Negative heat

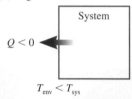

$Q < 0$

$T_{env} < T_{sys}$

(c) Thermal equilibrium

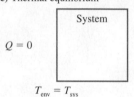

$Q = 0$

$T_{env} = T_{sys}$

(we'll look at the details in Chapter 18), energy is transferred when the *faster* molecules in the hotter object collide with the *slower* molecules in the cooler object. On average, these collisions cause the faster molecules to lose energy and the slower molecules to gain energy. The net result is that energy is transferred from the hotter object to the colder object. The process itself, whereby energy is transferred between the system and the environment via atomic-level collisions, is called a **thermal interaction.**

When you place a pan of water on the stove, heat is the energy transferred *from* the hotter flame *to* the cooler water. If you place the water in a freezer, heat is the energy transferred from the warmer water to the colder air in the freezer. A system is in **thermal equilibrium** with the environment, or two systems are in thermal equilibrium with each other, if there is no temperature difference.

It is worthwhile to compare this statement about heat and thermal interactions with the first paragraph about work in Section 17.2. The analogy would be complete if we were able to say that the environment (or an object in the environment) "does heat" on the system. Unfortunately, the English language doesn't work that way. Loosely speaking, we say that the environment "heats" the system.

Like work, **heat is not a state variable. That is, heat is not a property of the system.** Instead, heat is the amount of energy that moves between the system and the environment during a thermal interaction. It would not be meaningful to talk about a "change of heat." Thus heat appears in the energy equation simply as a value Q, never as ΔQ.

FIGURE 17.11 shows that Q is positive when energy is transferred *into* the system from the environment. This implies that $T_{env} > T_{sys}$. A negative Q represents heat transfer *from* the system to the environment when $T_{env} < T_{sys}$. The system is in thermal equilibrium with its environment when $T_{env} = T_{sys}$.

NOTE ▶ For both heat and work, a positive value indicates energy being transferred from the environment to the system. Table 17.1 summarizes the similarities and differences between work and heat. ◀

TABLE 17.1 Understanding work and heat

	Work	**Heat**
Interaction:	Mechanical	Thermal
Requires:	Force and displacement	Temperature difference
Process:	Macroscopic pushes and pulls	Microscopic collisions
Positive value:	$W > 0$ when a gas is compressed. Energy is transferred in.	$Q > 0$ when the environment is at a higher temperature than the system. Energy is transferred in.
Negative value:	$W < 0$ when a gas expands. Energy is transferred out.	$Q < 0$ when the system is at a higher temperature than the environment. Energy is transferred out.
Equilibrium:	A system is in mechanical equilibrium when there is no net force or torque on it.	A system is in thermal equilibrium when it is at the same temperature as the environment.

Units of Heat

Heat is energy transferred between the system and the environment. Consequently, the SI unit of heat is the joule. Historically, before the connection between heat and work had been recognized, a unit for measuring heat, the calorie, had been defined as

$$1 \text{ calorie} = 1 \text{ cal} = \text{the quantity of heat needed to change}$$
$$\text{the temperature of 1 g of water by 1 °C}$$

Once Joule established that heat is energy, it was apparent that the calorie is really a unit of energy. In today's SI units, the conversion is

$$1 \text{ cal} = 4.186 \text{ J}$$

The calorie you know in relation to food is not the same as the heat calorie. The *food calorie,* abbreviated Cal with a capital C, is

$$1 \text{ food calorie} = 1 \text{ Cal} = 1000 \text{ cal} = 1 \text{ kcal} = 4186 \text{ J}$$

The food calorie measures the food's chemical energy, stored energy that is available for doing work or for keeping your body warm. That extra dessert you ate last night containing 300 Cal has a chemical energy

$$E_{\text{chem}} = 300 \text{ Cal} = 3.00 \times 10^5 \text{ cal} = 1.26 \times 10^6 \text{ J}$$

We will not use calories in this textbook, but there are some fields of science and engineering where calories are still widely used. All the calculations you learn to do with joules can equally well be done with calories.

The Trouble with Heat

The trouble with heat is twofold: conceptual and linguistic. At the conceptual level, it is important to distinguish among *heat, temperature,* and *thermal energy.* These three ideas are related, but the distinctions between them are crucial. Common language can easily mislead you. If an object slides to a halt because of friction, most people say that the object's kinetic energy is "converted into heat." In fact, heat is not involved in this process. Nowhere was there a transfer of energy due to a temperature difference. Instead, the object's mechanical energy is transformed into the *thermal energy* of the atoms and molecules. In brief,

- Thermal energy is an energy *of the system* due to the motion of its atoms and molecules. It is a *form* of energy. Thermal energy is a state variable, and it makes sense to talk about how E_{th} changes during a process. The system's thermal energy continues to exist even if the system is isolated and not interacting thermally with its environment.
- Heat is energy transferred *between the system* and the environment as they interact. Heat is *not* a particular form of energy, nor is it a state variable. It makes no sense to talk about how heat changes. $Q = 0$ if a system does not interact thermally with its environment. Heat may cause the system's thermal energy to change, but that doesn't make heat and thermal energy the same.
- Temperature is a state variable that quantifies the "hotness" or "coldness" of a system. We haven't given a precise definition of temperature, but it is related to the thermal energy *per molecule.* A temperature difference is a requirement for a thermal interaction in which heat energy is transferred between the system and the environment.

It is especially important not to associate an observed temperature increase with heat. Heating a system is one way to change its temperature, but, as Joule showed, not the only way. You can also change the system's temperature by doing work on the system. **Observing the system tells us nothing about the process by which energy enters or leaves the system.**

We have two problems on the linguistic front. One, already alluded to, is those terms such as "heat flow" and "heat capacity" that are vestiges of history. These phrases, used even in scientific and technical discourse, incorrectly suggest that heat is a substance that can flow from one object to another or be contained in an object. With experience, scientists and engineers learn to use these phrases without meaning what the phrase, interpreted literally, seems to suggest.

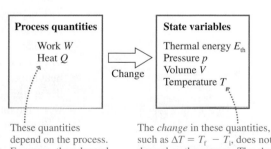

Process quantities and state variables.

A second problem is that the phrase "to heat" uses the word "heat" as a verb, whereas our definition of "heat" uses the word as a noun. These two uses make no distinction between the energy transferred and the process of transferring the energy. With work, the phrase "to *do* work" allows us to separate the process from the energy transferred (i.e., "the work") in the process.

Unfortunately, physics textbooks can't reinvent language. We will try to be very careful in our choice of words and phrases, and we will highlight points where the language is potentially confusing or misleading. Being forewarned will help you avoid some of these pitfalls.

STOP TO THINK 17.3   Which one or more of the following processes involves heat?

    a. The brakes in your car get hot when you stop.
    b. A steel block is held over a candle.
    c. You push a rigid cylinder of gas across a frictionless surface.
    d. You push a piston into a cylinder of gas, increasing the temperature of the gas.
    e. You place a cylinder of gas in hot water. The gas expands, causing a piston to rise and lift a weight. The temperature of the gas does not change.

17.4 The First Law of Thermodynamics

8.8–8.10

Heat was the missing piece that we needed to arrive at a completely general statement of the law of conservation of energy. Restating Equation 17.6,

$$\Delta E_{sys} = \Delta E_{mech} + \Delta E_{th} = W + Q$$

Work and heat, two ways of transferring energy between a system and the environment, cause the system's energy to change.

At this point in the text we are not interested in systems that have a macroscopic motion of the system as a whole. Moving macroscopic systems were important to us for many chapters, but now, as we investigate the thermal properties of a system, we would like the system as a whole to rest peacefully on the laboratory bench while we study it. So we will assume, throughout the remainder of Part IV, that $\Delta E_{mech} = 0$.

With this assumption clearly stated, the law of conservation of energy becomes

$$\Delta E_{th} = W + Q \qquad \text{(first law of thermodynamics)} \qquad (17.17)$$

The energy equation, in this form, is called the **first law of thermodynamics** or simply "the first law." The first law is a very general statement about the conservation of energy.

Chapters 10 and 11 introduced the basic energy model. It was called *basic* because it included work but not heat. The first law of thermodynamics has included heat, but it excludes situations where the mechanical energy changes. **FIGURE 17.12** is a pictorial representation of the **thermodynamic energy model** described by the first law. Work and heat are energies transferred between the system and the environment. Energy added to the system (W or Q positive) increases the system's thermal energy ($\Delta E_{th} > 0$). Likewise, the thermal energy decreases when energy is removed from the system.

Two comments are worthwhile:

1. The first law of thermodynamics doesn't tell us anything about the value of E_{th}, only how E_{th} changes. Doing 1 J of work changes the thermal energy by $\Delta E_{th} = 1$ J regardless of whether $E_{th} = 10$ J or 10,000 J.
2. The system's thermal energy isn't the only thing that changes. Work or heat that changes the thermal energy also changes the pressure, volume, temperature, and other state variables. The first law tells us only about ΔE_{th}. Other laws and relationships must be used to learn how the other state variables change.

FIGURE 17.12 The thermodynamic energy model.

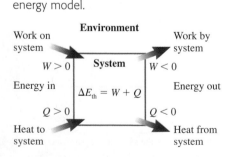

The first law is one of the most important analytic tools of thermodynamics. We'll use the first law in the remainder of this chapter to study some of the thermal properties of matter.

Three Special Ideal-Gas Processes

There are three ideal-gas processes in which one of the terms in the first law—ΔE_{th}, W, or Q—is zero. To investigate these processes, **FIGURE 17.13** shows a gas cylinder with three special properties:

- You can keep the gas volume from changing by inserting the locking pin into the piston. Without the pin, the piston can slide up or down. The piston is massless, frictionless, and insulated.
- You can change the gas pressure by adding or removing masses on top of the piston. Work is done as the piston moves the masses up and down.
- You can warm or cool the gas by placing the cylinder above a flame or on a block of ice. The thin bottom of the cylinder is the only surface through which heat energy can be transferred.

You learned in Chapter 16 (see Figure 16.12) that the gas pressure when the piston "floats" is determined by the atmospheric pressure and by the total mass M on the piston:

$$p_{gas} = p_{atmos} + \frac{Mg}{A} \qquad (17.18)$$

The pressure doesn't change as the piston moves unless you change the mass. This is a particularly important point to understand. Equation 17.18 is *not* valid when the piston is locked. The pressure with the piston locked could be either higher or lower than Equation 17.18.

An isochoric cooling process ($W = 0$): No work is done in an isochoric (constant volume) process because the piston doesn't move. To cool the gas without doing work:

- Insert the locking pin so that the volume cannot change.
- Place the cylinder on the block of ice. Heat energy will be transferred from the gas to the ice, causing the gas temperature and pressure to fall.
- Remove the cylinder from the ice when the desired pressure is reached.
- Remove masses from the piston until the total mass M balances the new gas pressure. This step must be done before removing the locking pin; otherwise, the piston will move when the pin is removed.
- Remove the locking pin.

Figure 17.7a showed the pV diagram. The final point is on a lower isotherm than the initial point, so $T_f < T_i$. No work was done, but heat energy was transferred out of the gas ($Q < 0$) and the thermal energy of the gas decreased ($\Delta E_{th} < 0$) as the temperature fell. **FIGURE 17.14** shows this result on a first-law bar chart. We don't know the value of the initial thermal energy $E_{th\,i}$ so the height of the $E_{th\,i}$ bar is arbitrary. Even so, we see that the thermal energy has decreased by the amount of energy that left the system as heat.

An isothermal expansion ($\Delta E_{th} = 0$): The thermal energy does not change in an isothermal process because the temperature of the gas doesn't change. To expand the gas without changing its thermal energy:

- Place the cylinder over the flame. Heat energy will be transferred to the gas, and the gas will begin to expand.
- The product pV must remain constant during an isothermal process. Slowly remove masses from the piston to reduce the pressure as the volume increases. The temperature remains constant as heat energy from the flame balances the negative work done on the gas as it expands.
- Remove the cylinder from the flame when the gas reaches the desired volume.

Figure 17.8 showed the pV diagram. $\Delta E_{th} = 0$ in an isothermal process ($\Delta T = 0$), so the first law $\Delta E_{th} = W + Q$ can be satisfied only if $W = -Q$. Heat energy is transferred to the gas, but the temperature of the gas doesn't change. Instead, the energy

FIGURE 17.13 The gas can be heated and have work done on it.

Masses determine the gas pressure. Work is done as the masses move up and down.

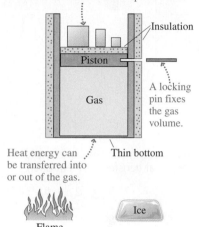

Heat energy can be transferred into or out of the gas.

FIGURE 17.14 A first-law bar chart for a process that does no work.

$$E_{th\,i} + W + Q = E_{th\,f}$$

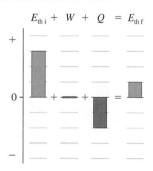

FIGURE 17.15 A first-law bar chart for a
process that doesn't change the thermal
energy.

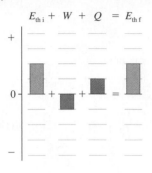

FIGURE 17.15 A first-law bar chart for a
process that doesn't change the thermal
energy.

causes the gas to expand and do the work of lifting the masses. The work done *on* the
gas by the piston is negative as the gas expands. This information is shown on the
first-law bar chart of **FIGURE 17.15**.

> **NOTE** ▶ It is surprising, but true, that we can heat the system without changing its
> temperature. But to do so, we must have a process in which the energy coming into
> the system as heat is exactly balanced by the energy leaving the system as work.
> **The important point is that $\Delta T = 0$ does *not* mean $Q = 0$.** ◀

An adiabatic compression ($Q = 0$): A process in which no heat energy is transferred between the system and the environment is called an **adiabatic process.**
Although the system cannot have thermal interactions with its environment, it can still
have mechanical interactions as the insulated piston pushes or pulls on the gas. To
compress the gas without heat:

- Add insulation beneath the cylinder.
- Slowly add masses to the piston, increasing the pressure. The piston will slowly
  descend, compressing the gas and decreasing its volume.
- Stop adding masses when the gas reaches the desired volume.

FIGURE 17.16 A first-law bar chart for a
process that transfers no heat energy.

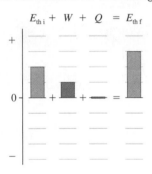

$Q = 0$ in an adiabatic process, so the first law $\Delta E_{th} = W + Q$ can be satisfied only if
$\Delta E_{th} = W$. Work is done on the gas to compress it. The energy transferred into the system as work increases the thermal energy—and thus the temperature—of the gas. This
information is shown on the first-law bar chart of **FIGURE 17.16**.

> **NOTE** ▶ Just because the system is well insulated—thermally isolated from the
> environment—does not mean its temperature remains constant. Energy coming
> into the system as work has the same consequences as if the energy entered the sys-
> tem as heat. An adiabatic compression uses work to increase the temperature of the
> gas. Similarly, an adiabatic expansion lowers the temperature of the gas. **The
> important point is that $Q = 0$ does *not* mean $\Delta T = 0$.** ◀

We'll examine adiabatic gas processes and their pV curve later in the chapter. For
now, make sure you understand which quantities are zero and which aren't in these
three special processes.

STOP TO THINK 17.4   Which first-law bar chart describes the process shown in the pV diagram?

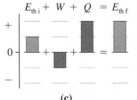

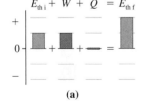

(a)

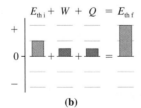

(b)

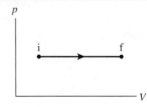

(c)

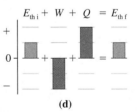

(d)

17.5 Thermal Properties of Matter

Joule established that heat and work are energy transferred between a system and its
environment. Heat and work are equivalent in the sense that the change of the system is
exactly the same whether you transfer heat energy to it or do an equal amount of work
on it. Adding energy to the system, or removing it, changes the system's thermal energy.

What happens to a system when you change its thermal energy? In this section we'll consider two distinct possibilities:

- The temperature of the system changes.
- The system undergoes a phase change, such as melting or freezing.

Temperature Change and Specific Heat

Suppose you do an experiment in which you add energy to water, either by doing work on it or by transferring heat to it. Either way, you will find that adding 4190 J of energy raises the temperature of 1 kg of water by 1 K. If you were fortunate enough to have 1 kg of gold, you would need to add only 129 J of energy to raise its temperature by 1 K.

The amount of energy that raises the temperature of 1 kg of a substance by 1 K is called the **specific heat** of that substance. The symbol for specific heat is c. Water has specific heat $c_{water} = 4190$ J/kg K. The specific heat of gold is $c_{gold} = 129$ J/kg K. Specific heat depends only on the material from which an object is made. Table 17.2 provides some specific heats for common liquids and solids.

NOTE ▶ The term *specific heat* does not use the word "heat" in the way that we have defined it. Specific heat is an old idea, dating back to the days of the caloric theory when heat was thought to be a substance contained in the object. The term has continued in use even though our understanding of heat has changed. ◀

If energy c is required to raise the temperature of 1 kg of substance by 1 K, then energy Mc is needed to raise the temperature of mass M by 1 K and $(Mc)\Delta T$ is needed to raise the temperature of mass M by ΔT. In other words, the thermal energy of the system changes by

$$\Delta E_{th} = Mc\Delta T \qquad \text{(temperature change)} \qquad (17.19)$$

when its temperature changes by ΔT. ΔE_{th} can be either positive (thermal energy increases as the temperature goes up) or negative (thermal energy decreases as the temperature goes down). Recall that uppercase M is used for the mass of an entire system while lowercase m is reserved for the mass of an atom or molecule.

NOTE ▶ In practice, ΔT is usually measured in °C. But the Kelvin and the Celsius temperature scales have the same step size, so ΔT in K has exactly the same numerical value as ΔT in °C. Thus

- You do not need to convert temperatures from °C to K if you need only a temperature *change* ΔT.
- You do need to convert anytime you need the actual temperature T. ◀

The first law of thermodynamics, $\Delta E_{th} = W + Q$, allows us to write Equation 17.19 as $Mc\Delta T = W + Q$. In other words, **we can change the system's temperature either by heating it or by doing an equivalent amount of work on it.** In working with solids and liquids, we almost always change the temperature by heating. If $W = 0$, which we will assume for the rest of this section, then the heat needed to bring about a temperature change ΔT is

$$Q = Mc\Delta T \qquad \text{(temperature change)} \qquad (17.20)$$

Because $\Delta T = \Delta E_{th}/Mc$, it takes more energy to change the temperature of a substance with a large specific heat than to change the temperature of a substance with a small specific heat. You can think of specific heat as measuring the *thermal inertia* of a substance. Metals, with small specific heats, warm up and cool down quickly. A piece of aluminum foil can be safely held within seconds of removing it from a hot oven. Water, with a very large specific heat, is slow to warm up and slow to cool down. This is fortunate for us. The large thermal inertia of water is essential for the biological processes of life. We wouldn't be here studying physics if water had a small specific heat!

TABLE 17.2 Specific heats and molar specific heats of solids and liquids

Substance	c (J/kg K)	C (J/mol K)
Solids		
Aluminum	900	24.3
Copper	385	24.4
Iron	449	25.1
Gold	129	25.4
Lead	128	26.5
Ice	2090	37.6
Liquids		
Ethyl alcohol	2400	110.4
Mercury	140	28.1
Water	4190	75.4

EXAMPLE 17.3 Quenching hot aluminum in ethyl alcohol

A 50.0 g aluminum disk at 300°C is placed in 200 cm³ of ethyl alcohol at 10.0°C, then quickly removed. The aluminum temperature is found to have dropped to 120°C. What is the new temperature of the ethyl alcohol?

MODEL Heat is the energy transferred due to a temperature difference. If we assume that the container holding the alcohol is well insulated, then the disk and the alcohol interact with each other but nothing else. Conservation of energy tells us that the heat energy transferred out of the disk is the heat energy transferred into the alcohol.

SOLVE The temperature change of the disk is $\Delta T_{Al} = (120°C - 300°C) = -180°C = -180$ K. It is negative because the temperature decreases. The energy removed from the disk is

$$Q_{Al} = Mc\Delta T = (0.0500 \text{ kg})(900 \text{ J/kg K})(-180 \text{ K}) = -8100 \text{ J}$$

Q_{Al} is negative because the energy is transferred out of the aluminum. The ethyl alcohol *gains* 8100 J of energy; thus $Q_{ethyl} = +8100$ J. We need to know the mass of the ethyl alcohol. Its density was given in Table 16.1 as $\rho = 790$ kg/m³; hence its mass is

$$M = \rho V = (790 \text{ kg/m}^3)(200 \times 10^{-6} \text{ m}^3) = 0.158 \text{ kg}$$

The heat from the aluminum causes the alcohol's temperature to change by

$$\Delta T = \frac{Q_{ethyl}}{Mc} = \frac{8100 \text{ J}}{(0.158 \text{ kg})(2400 \text{ J/kg K})} = 21.4 \text{ K}$$
$$= 21.4°C$$

The ethyl alcohol ends up at temperature

$$T_f = T_i + \Delta T = 10.0°C + 21.4°C = 31.4°C$$

The **molar specific heat** is the amount of energy that raises the temperature of 1 mol of a substance by 1 K. We'll use an uppercase C for the molar specific heat. The heat needed to bring about a temperature change ΔT of n moles of substance is

$$Q = nC\Delta T \tag{17.21}$$

Molar specific heats are listed in Table 17.2. Look at the five elemental solids (excluding ice). All have C very near 25 J/mol K. If we were to expand the table, we would find that most elemental solids have $C \approx 25$ J/mol K. This can't be a coincidence, but what is it telling us? This is a puzzle we will address in Chapter 18, where we will explore thermal energy at the atomic level.

Phase Change and Heat of Transformation

Suppose you start with a system in its solid phase and heat it at a steady rate. **FIGURE 17.17**, which you saw in Chapter 16, shows how the system's temperature changes. At first, the temperature increases linearly. This is not hard to understand because Equation 17.20 can be written

$$\text{slope of the } T\text{-versus-}Q \text{ graph} = \frac{\Delta T}{Q} = \frac{1}{Mc} \tag{17.22}$$

The slope of the graph depends inversely on the system's specific heat. A constant specific heat implies a constant slope and thus a linear graph. In fact, you can measure c from such a graph.

NOTE ▶ The different slopes indicate that the solid, liquid, and gas phases of a substance have different specific heats. ◀

But there are times, shown as horizontal line segments, during which heat is being transferred to the system but the temperature isn't changing. These are *phase changes*. The thermal energy continues to increase during a phase change, but the additional energy goes into breaking molecular bonds rather than speeding up the molecules. **A phase change is characterized by a change in thermal energy without a change in temperature.**

The amount of heat energy that causes 1 kg of a substance to undergo a phase change is called the **heat of transformation** of that substance. For example, laboratory experiments show that 333,000 J of heat are needed to melt 1 kg of ice at 0°C. The symbol for heat of transformation is L. The heat required for the entire system of mass M to undergo a phase change is

$$Q = ML \quad \text{(phase change)} \tag{17.23}$$

FIGURE 17.17 The temperature of a system that is heated at a steady rate.

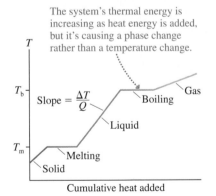

Cumulative heat added

Heat of transformation is a generic term that refers to any phase change. Two specific heats of transformation are the **heat of fusion** L_f, the heat of transformation between a solid and a liquid, and the **heat of vaporization** L_v, the heat of transformation between a liquid and a gas. The heat needed for these phase changes is

$$Q = \begin{cases} \pm ML_f & \text{melt/freeze} \\ \pm ML_v & \text{boil/condense} \end{cases} \quad (17.24)$$

where the $\pm$ indicates that heat must be *added* to the system during melting or boiling but *removed* from the system during freezing or condensing. **You must explicitly include the minus sign when it is needed.**

Table 17.3 gives the heats of transformation of a few substances. Notice that the heat of vaporization is always much larger than the heat of fusion. We can understand this. Melting breaks just enough molecular bonds to allow the system to lose rigidity and flow. Even so, the molecules in a liquid remain close together and loosely bonded. Vaporization breaks all bonds completely and sends the molecules flying apart. This process requires a larger increase in the thermal energy and thus a larger quantity of heat.

Lava—molten rock—undergoes a phase change when it contacts the much colder water. This is one way in which new islands are formed.

TABLE 17.3 Melting/boiling temperatures and heats of transformation

Substance	T_m (°C)	L_f (J/kg)	T_b (°C)	L_v (J/kg)
Nitrogen (N$_2$)	−210	0.26×10^5	−196	1.99×10^5
Ethyl alcohol	−114	1.09×10^5	78	8.79×10^5
Mercury	−39	0.11×10^5	357	2.96×10^5
Water	0	3.33×10^5	100	22.6×10^5
Lead	328	0.25×10^5	1750	8.58×10^5

EXAMPLE 17.4 Turning ice into steam
How much heat is required to change 200 mL of ice at −20°C (a typical freezer temperature) into steam?

MODEL Changing ice to steam requires four steps: Raise the temperature of the ice to 0°C, melt the ice to liquid water at 0°C, raise the water temperature to 100°C, then boil the water to produce steam at 100°C.

SOLVE The mass is $M = \rho V$. The density of ice (from Table 16.1) is 920 kg/m³, and $V = 200$ mL $= 200$ cm³ $= 2.00 \times 10^{-4}$ m³. Thus

$$M = \rho V = (920 \text{ kg/m}^3)(2.00 \times 10^{-4} \text{ m}^3) = 0.184 \text{ kg}$$

The heat needed for each step is

$$Q_1 = Mc_{ice}\Delta T_{ice} = (0.184 \text{ kg})(2090 \text{ J/kg K})(20 \text{ K}) = 7,700 \text{ J}$$

$$Q_2 = ML_f = (0.184 \text{ kg})(3.33 \times 10^5 \text{ J/kg}) = 61,300 \text{ J}$$

$$Q_3 = Mc_{water}\Delta T_{water} = (0.184 \text{ kg})(4190 \text{ J/kg K})(100 \text{ K}) = 77,100 \text{ J}$$

$$Q_4 = ML_v = (0.184 \text{ kg})(22.6 \times 10^5 \text{ J/kg}) = 415,800 \text{ J}$$

NOTE ▶ We used the specific heat of ice while warming the system in its solid phase. Then we used the specific heat of *water* while warming the system in its liquid phase. ◀

The total heat required is

$$Q = Q_1 + Q_2 + Q_3 + Q_4 = 562,000 \text{ J}$$

ASSESS Roughly 75% of the heat is used to change the water from a 100°C liquid to a 100°C gas. This is consistent with your experience that it takes much longer for a pan of water to boil away than it does to reach boiling.

STOP TO THINK 17.5 Objects A and B are brought into close thermal contact with each other, but they are well isolated from their surroundings. Initially $T_A = 0$°C and $T_B = 100$°C. The specific heat of A is less than the specific heat of B. The two objects will soon reach a common final temperature T_f. The final temperature is

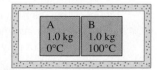

a. $T_f > 50$°C   b. $T_f = 50$°C   c. $T_f < 50$°C

17.6 Calorimetry

At one time or another you've probably put an ice cube into a hot drink to cool it quickly. You were engaged, in a somewhat trial-and-error way, in a practical aspect of heat transfer known as **calorimetry.**

FIGURE 17.18 shows two systems thermally interacting with each other but isolated from everything else. Suppose they start at different temperatures T_1 and T_2. As you know from experience, heat energy will be transferred from the hotter to the colder system until they reach a common final temperature T_f. The systems will then be in thermal equilibrium and the temperature will not change further.

The insulation prevents any heat energy from being transferred to or from the environment, so energy conservation tells us that any energy leaving the hotter system must enter the colder system. That is, the systems *exchange* energy with no net loss or gain. The concept is straightforward, but to state the idea mathematically we need to be careful with signs.

Let Q_1 be the energy transferred to system 1 as heat. Similarly, Q_2 is the energy transferred to system 2. The fact that the systems are merely exchanging energy can be written $|Q_1| = |Q_2|$. That is, the energy *lost* by the hotter system is the energy *gained* by the colder system. Thus Q_1 and Q_2 have opposite signs: $Q_1 = -Q_2$. No energy is exchanged with the environment, hence it makes more sense to write this relationship as

$$Q_{net} = Q_1 + Q_2 = 0 \qquad (17.25)$$

This idea is not limited to the interaction of only two systems. If three or more systems are combined in isolation from the rest of their environment, each at a different initial temperature, they will all come to a common final temperature that can be found from the relationship

$$Q_{net} = Q_1 + Q_2 + Q_3 + \cdots = 0 \qquad (17.26)$$

NOTE ▶ The signs are very important in calorimetry problems. ΔT is always $T_f - T_i$, so ΔT and Q are negative for any system whose temperature decreases. The proper sign of Q for any phase change must be supplied *by you,* depending on the direction of the phase change. ◀

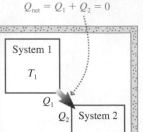

FIGURE 17.18 Two systems interact thermally.

Heat energy is transferred from system 1 to system 2. Energy conservation requires

$$|Q_1| = |Q_2|$$

Opposite signs mean that

$$Q_{net} = Q_1 + Q_2 = 0$$

System 1
T_1
Q_1
Q_2 System 2
T_2

PROBLEM-SOLVING
STRATEGY 17.2   **Calorimetry problems**

MODEL Identify the interacting systems. Assume that they are isolated from the larger environment.

VISUALIZE List known information and identify what you need to find. Convert all quantities to SI units.

SOLVE The mathematical representation, which is a statement of energy conservation, is

$$Q_{net} = Q_1 + Q_2 + \cdots = 0$$

- For systems that undergo a temperature change, $Q = Mc(T_f - T_i)$. Be sure to have the temperatures T_i and T_f in the correct order.
- For systems that undergo a phase change, $Q = \pm ML$. Supply the correct sign by observing whether energy enters or leaves the system.
- Some systems may undergo a temperature change *and* a phase change. Treat the changes separately. The heat energy is $Q = Q_{\Delta T} + Q_{phase}$.

ASSESS Is the final temperature in the middle? T_f that is higher or lower than all initial temperatures is an indication that something is wrong, usually a sign error.

NOTE ► You may have learned to solve calorimetry problems in other courses by writing $Q_{gained} = Q_{lost}$. That is, by balancing heat gained with heat lost. That approach works in simple problems, but it has two drawbacks. First, you often have to "fudge" the signs to make them work. Second, and more serious, you can't extend this approach to a problem with three or more interacting systems. Using $Q_{net} = 0$ is much preferred. ◄

EXAMPLE 17.5 Calorimetry with a phase change

Your 500 mL soda is at 20°C, room temperature, so you add 100 g of ice from the −20°C freezer. Does all the ice melt? If so, what is the final temperature? If not, what fraction of the ice melts? Assume that you have a well-insulated cup.

MODEL We have a thermal interaction between the soda, which is essentially water, and the ice. We need to distinguish between three possible outcomes. If all the ice melts, then $T_f > 0°C$. It's also possible that the soda will cool to 0°C before all the ice has melted, leaving the ice and liquid in equilibrium at 0°C. A third possibility is that the soda will freeze solid before the ice warms up to 0°C. That seems unlikely here, but there are situations, such as the pouring of molten metal out of furnaces, when all the liquid does solidify. We need to distinguish between these before knowing how to proceed.

VISUALIZE All the initial temperatures, masses, and specific heats are known. The final temperature of the combined soda + ice system is unknown.

SOLVE Let's first calculate the heat needed to melt all the ice and leave it as liquid water at 0°C. To do so, we must warm the ice to 0°C, then change it to water. The heat input for this two-stage process is

$$Q_{melt} = M_i c_i (20 \text{ K}) + M_i L_f = 37,500 \text{ J}$$

where L_f is the heat of fusion of water. It is used as a *positive* quantity because we must *add* heat to melt the ice. Next, let's calculate how much heat energy will leave the soda if it cools all the

way to 0°C. The volume is $V = 500 \text{ mL} = 5.00 \times 10^{-4} \text{ m}^3$ and thus the mass is $M_s = \rho V = 0.500 \text{ kg}$. The heat is

$$Q_{cool} = M_s c_w (-20 \text{ K}) = -41,900 \text{ J}$$

where $\Delta T = -20 \text{ K}$ because the temperature decreases. Because $|Q_{cool}| > Q_{melt}$, the soda has sufficient energy to melt all the ice. Hence the final state will be all liquid at $T_f > 0$. (Had we found $|Q_{cool}| < Q_{melt}$, then the final state would have been an ice-liquid mixture at 0°C.)

Energy conservation requires $Q_{ice} + Q_{soda} = 0$. The heat Q_{ice} consists of three terms: warming the ice to 0°C, melting the ice to water at 0°C, then warming the 0°C water to T_f. The mass will still be M_i in the last of these steps because it is the "ice system," but we need to use the specific heat of *liquid water*. Thus

$$Q_{ice} + Q_{soda} = \left[M_i c_i (20 \text{ K}) + M_i L_f + M_i c_w (T_f - 0°C) \right]$$
$$+ M_s c_w (T_f - 20°C) = 0$$

We've already done part of the calculation, allowing us to write

$$37,500 \text{ J} + M_i c_w (T_f - 0°C) + M_s c_w (T_f - 20°C) = 0$$

Solving for T_f gives

$$T_f = \frac{20 M_s c_w - 37,500}{M_i c_w + M_s c_w} = 1.7°C$$

ASSESS As expected, the soda has been cooled to nearly the freezing point.

EXAMPLE 17.6 Three interacting systems

A 200 g piece of iron at 120°C and a 150 g piece of copper at −50°C are dropped into an insulated beaker containing 300 g of ethyl alcohol at 20°C. What is the final temperature?

MODEL Here you can't use a simple $Q_{gained} = Q_{lost}$ approach because you don't know whether the alcohol is going to warm up or cool down.

VISUALIZE All the initial temperatures, masses, and specific heats are known. We need to find the final temperature.

SOLVE Energy conservation requires

$$Q_i + Q_c + Q_e = M_i c_i (T_f - 120°C) + M_c c_c (T_f - (-50°C))$$
$$+ M_e c_e (T_f - 20°C) = 0$$

Solving for T_f gives

$$T_f = \frac{120 M_i c_i - 50 M_c c_c + 20 M_e c_e}{M_i c_i + M_c c_c + M_e c_e} = 25.7°C$$

ASSESS The temperature is between the initial iron and copper temperatures, as expected. It turns out that the alcohol warms up ($Q_e > 0$), but we had no way to know this without doing the calculation.

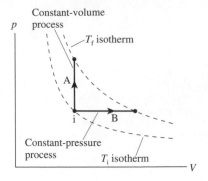

FIGURE 17.19 Processes A and B have the same ΔT and the same ΔE_{th}, but they require different amounts of heat.

17.7 The Specific Heats of Gases

Specific heats are given in Table 17.2 for solids and liquids. Gases are harder to characterize because the heat required to cause a specified temperature change depends on the *process* by which the gas changes state.

FIGURE 17.19 shows two isotherms on the pV diagram for a gas. Processes A and B, which start on the T_i isotherm and end on the T_f isotherm, have the *same* temperature change $\Delta T = T_f - T_i$. But process A, which takes place at constant volume, requires a *different* amount of heat than does process B, which occurs at constant pressure. The reason is that work is done in process B but not in process A. This is a situation that we are now equipped to analyze.

It is useful to define two different versions of the specific heat of gases, one for constant-volume (isochoric) processes and one for constant-pressure (isobaric) processes. We will define these as molar specific heats because we usually do gas calculations using moles instead of mass. The quantity of heat needed to change the temperature of n moles of gas by ΔT is

$$Q = nC_V\Delta T \qquad \text{(temperature change at constant volume)}$$

$$Q = nC_P\Delta T \qquad \text{(temperature change at constant pressure)}$$

(17.27)

where C_V is the **molar specific heat at constant volume** and C_P is the **molar specific heat at constant pressure.** Table 17.4 gives the values of C_V and C_P for a few common monatomic and diatomic gases. The units are J/mol K.

NOTE ▶ Equation 17.27 applies to two specific ideal-gas processes. In a general gas process, for which neither p nor V is constant, we have no direct way to relate Q to ΔT. In that case, the heat must be found indirectly from the first law as $Q = \Delta E_{th} - W$. ◄

TABLE 17.4 Molar specific heats of gases (J/mol K)

Gas	C_P	C_V	$C_P - C_V$
Monatomic Gases			
He	20.8	12.5	8.3
Ne	20.8	12.5	8.3
Ar	20.8	12.5	8.3
Diatomic Gases			
H_2	28.7	20.4	8.3
N_2	29.1	20.8	8.3
O_2	29.2	20.9	8.3

EXAMPLE 17.7 Heating and cooling a gas
Three moles of O_2 gas are at 20.0°C. 600 J of heat energy are transferred to the gas at constant pressure, then 600 J are removed at constant volume. What is the final temperature? Show the process on a pV diagram.

MODEL O_2 is a diatomic ideal gas. The gas is heated as an isobaric process, then cooled as an isochoric process.

SOLVE The heat transferred during the constant-pressure process causes a temperature rise

$$\Delta T = T_2 - T_1 = \frac{Q}{nC_P} = \frac{600 \text{ J}}{(3.0 \text{ mol})(29.2 \text{ J/mol K})} = 6.8°C$$

where C_P for oxygen was taken from Table 17.4. Heating leaves the gas at temperature $T_2 = T_1 + \Delta T = 26.8°C$. The temperature then falls as heat is removed during the constant-volume process:

$$\Delta T = T_3 - T_2 = \frac{Q}{nC_V} = \frac{-600 \text{ J}}{(3.0 \text{ mol})(20.9 \text{ J/mol K})} = -9.5°C$$

We used a *negative* value for Q because heat energy is transferred from the gas to the environment. The final temperature of the gas

is $T_3 = T_2 + \Delta T = 17.3°C$. **FIGURE 17.20** shows the process on a pV diagram. The gas expands (moves horizontally on the diagram) as heat is added, then cools at constant volume (moves vertically on the diagram) as heat is removed.

FIGURE 17.20 The pV diagram for Example 17.7.

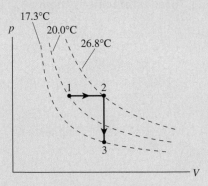

ASSESS The final temperature is lower than the initial temperature because $C_P > C_V$.

EXAMPLE 17.8 Calorimetry with a gas and a solid
The interior volume of a 200 g hollow aluminum box is 800 cm³. The box contains nitrogen gas at STP. A 20 cm³ block of copper at a temperature of 300°C is placed inside the box, then the box is sealed. What is the final temperature?

MODEL This example has three interacting systems: the aluminum box, the nitrogen gas, and the copper block. They must all come to a common final temperature T_f.

VISUALIZE The box and gas have the same initial temperature: $T_{Al} = T_{N2} = 0°C$. The box doesn't change size, so this is a constant-volume process. The final temperature is unknown.

SOLVE Although one of the systems is now a gas, the calorimetry equation $Q_{net} = Q_{Al} + Q_{N2} + Q_{Cu} = 0$ is still appropriate. In this case,

$$Q_{net} = m_{Al}c_{Al}(T_f - T_{Al}) + n_{N2}C_V(T_f - T_{N2})$$
$$+ m_{Cu}c_{Cu}(T_f - T_{Cu}) = 0$$

Notice that we used masses and specific heats for the solids but moles and the molar specific heat for the gas. We used C_V because this is a constant-volume process. Solving for T_f gives

$$T_f = \frac{m_{Al}c_{Al}T_{Al} + n_{N2}C_VT_{N2} + m_{Cu}c_{Cu}T_{Cu}}{m_{Al}c_{Al} + n_{N2}C_V + m_{Cu}c_{Cu}}$$

The specific heat values are found in Tables 17.2 and 17.4. The mass of the copper is

$$m_{Cu} = \rho_{Cu}V_{Cu} = (8920 \text{ kg/cm}^3)(20 \times 10^{-6} \text{ m}^3) = 0.178 \text{ kg}$$

The number of moles of the gas is found from the ideal-gas law, using the initial conditions. Notice that inserting the copper block *displaces* 20 cm³ of gas; hence the gas volume is only $V = 780 \text{ cm}^3 = 7.80 \times 10^{-4} \text{ m}^3$. Thus

$$n_{N2} = \frac{pV}{RT} = 0.0348 \text{ mol}$$

Computing the final temperature gives $T_f = 83°C$.

C_P and C_V

You may have noticed two curious features in Table 17.4. First, the molar specific heats of monatomic gases are *all alike*. And the molar specific heats of diatomic gases, while different from monatomic gases, are again *very nearly alike*. We saw a similar feature in Table 17.2 for the molar specific heats of solids. Second, the *difference* $C_P - C_V = 8.3 \text{ J/mol K}$ is the same in every case. And, most puzzling of all, the value of $C_P - C_V$ appears to be equal to the universal gas constant R! Why should this be?

**Actⁱv
Physⁱcs** 8.7

The relationship between C_V and C_P hinges on one crucial idea: ΔE_{th}, **the change in the thermal energy of a gas, is the same for *any* two processes that have the same ΔT**. The thermal energy of a gas is associated with temperature, so any process that changes the gas temperature from T_i to T_f has the same ΔE_{th} as any other process that goes from T_i to T_f. Furthermore, the first law $\Delta E_{th} = Q + W$ tells us that a gas cannot distinguish between heat and work. The system's thermal energy changes in response to energy added to or removed from the system, but the response of the gas is the same whether you heat the system, do work on the system, or do some combination of both. Thus **any two processes that change the thermal energy of the gas by ΔE_{th}, will cause the same temperature change ΔT.**

With that in mind, look back at Figure 17.19. Both gas processes have the same ΔT, so both have the same value of ΔE_{th}. Process A is an isochoric process in which no work is done (the piston doesn't move), so the first law for this process is

$$(\Delta E_{th})_A = W + Q = 0 + Q_{const vol} = nC_V\Delta T \tag{17.28}$$

Process B is an isobaric process. You learned earlier that the work done on the gas during an isobaric process is $W = -p\Delta V$. Thus

$$(\Delta E_{th})_B = W + Q = -p\Delta V + Q_{const press} = -p\Delta V + nC_P\Delta T \tag{17.29}$$

$(\Delta E_{th})_B = (\Delta E_{th})_A$ because both have the same ΔT, so we can equate the right sides of Equations 17.28 and 17.29:

$$-p\Delta V + nC_P\Delta T = nC_V\Delta T \tag{17.30}$$

For the final step, we can use the ideal-gas law $pV = nRT$ to relate ΔV and ΔT during process B. For any gas process,

$$\Delta(pV) = \Delta(nRT) \tag{17.31}$$

For a constant-pressure process, where p is constant, Equation 17.31 becomes

$$p\Delta V = nR\Delta T \qquad (17.32)$$

Substituting this expression for $p\Delta V$ into Equation 17.30 gives

$$-nR\Delta T + nC_P\Delta T = nC_V\Delta T \qquad (17.33)$$

The $n\Delta T$ cancels, and we are left with

$$C_P = C_V + R \qquad (17.34)$$

This result, which applies to all ideal gases, is exactly what we see in the data of Table 17.4.

But that's not the only conclusion we can draw. Equation 17.28 found that $\Delta E_{th} = nC_V\Delta T$ for a constant-volume process. But we had just noted that ΔE_{th} is the same for *all* gas processes that have the same ΔT. Consequently, this expression for ΔE_{th} is equally true for any other process. That is

$$\Delta E_{th} = nC_V\Delta T \qquad \text{(any ideal-gas process)} \qquad (17.35)$$

Compare this result to Equation 17.27. We first made a distinction between constant-volume and constant-pressure processes, but now we're saying that Equation 17.35 is true for any process. Are we contradicting ourselves? No, the difference lies in what you need to calculate.

- The change in thermal energy when the temperature changes by ΔT is the same for any process. That's Equation 17.35.
- The *heat* required to bring about the temperature change depends on what the process is. That's Equation 17.27. An isobaric process requires more heat than an isochoric process that produces the same ΔT.

The reason for the difference is seen by writing the first law as $Q = \Delta E_{th} - W$. In an isochoric process, where $W = 0$, *all* the heat input is used to increase the gas temperature. But in an isobaric process, some of the energy that enters the system as heat then leaves the system as work ($W < 0$) done by the expanding gas. Thus more heat is needed to produce the same ΔT.

Heat Depends on the Path

FIGURE 17.21 Is the heat input along these two paths the same or different?

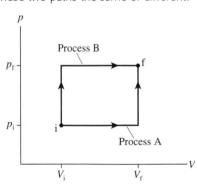

Consider the two ideal-gas processes shown in **FIGURE 17.21**. Even though the initial and final states are the same, the heat added during these two processes is *not* the same. We can use the first law $\Delta E_{th} = W + Q$ to see why.

The thermal energy is a state variable. That is, its value depends on the state of the gas, not the process by which the gas arrived at that state. Thus $\Delta E_{th} = E_{th\,f} - E_{th\,i}$ is the same for both processes. If ΔE_{th} is the same for processes A and B, then $W_A + Q_A = W_B + Q_B$.

You learned in Section 17.2 that the work done during an ideal-gas process depends on the path in the pV diagram. There's more area under the process B curve, so $|W_B| > |W_A|$. Both values of W are negative because the gas expands, so W_B is more negative than W_A. Consequently, $W_A + Q_A$ can equal $W_B + Q_B$ only if $Q_B > Q_A$. **The heat added or removed during an ideal-gas process depends on the path followed through the pV diagram.**

Adiabatic Processes

Section 17.4 introduced the idea of an *adiabatic process,* a process in which no heat energy is transferred ($Q = 0$). **FIGURE 17.22** on the next page compares an adiabatic process with isothermal and isochoric processes. We're now prepared to look at adiabatic processes in more detail.

In practice, there are two ways that an adiabatic process can come about. First, a gas cylinder can be completely surrounded by thermal insulation, such as thick pieces of Styrofoam. The environment can interact mechanically with the gas by pushing or pulling on the insulated piston, but there is no thermal interaction.

Second, the gas can be expanded or compressed very rapidly in what we call an *adiabatic expansion* or an *adiabatic compression*. In a rapid process there is essentially no time for heat to be transferred between the gas and the environment. We've already alluded to the idea that heat is transferred via atomic-level collisions. These collisions take time. If you stick one end of a copper rod into a flame, the other end will eventually get too hot to hold—but not instantly. Some amount of time is required for heat to be transferred from one end to the other. A process that takes place faster than the heat can be transferred is adiabatic.

NOTE ▶ You may recall reading in Chapter 16 that we are going to study only quasi-static processes, processes that proceed slowly enough to remain essentially in equilibrium at all times. Now we're proposing to study processes that take place very rapidly. Isn't this a contradiction? Yes, to some extent it is. What we need to establish are the appropriate time scales. How slow must a process go to be quasi-static? How fast must it go to be adiabatic? These types of calculations must be deferred to a more advanced course. It turns out—fortunately!—that many practical applications, such as the compression strokes in gasoline and diesel engines, are fast enough to be adiabatic yet slow enough to still be considered quasi-static. ◀

For an adiabatic process, with $Q = 0$, the first law of thermodynamics is $\Delta E_{th} = W$. Compressing a gas adiabatically ($W > 0$) increases the thermal energy. Thus **an adiabatic compression raises the temperature of a gas.** A gas that expands adiabatically ($W < 0$) gets colder as its thermal energy decreases. Thus **an adiabatic expansion lowers the temperature of a gas.** You can use an adiabatic process to change the gas temperature without using heat!

The work done in an adiabatic process goes entirely to changing the thermal energy of the gas. But we just found that $\Delta E_{th} = nC_V\Delta T$ for *any* process. Thus

$$W = nC_V\Delta T \qquad \text{(adiabatic process)} \qquad (17.36)$$

Equation 17.36 joins with the equations we derived earlier for the work done in isochoric, isobaric, and isothermal processes.

Gas processes can be represented as trajectories in the pV diagram. For example, a gas moves along a hyperbola during an isothermal process. How does an adiabatic process appear in a pV diagram? The result is more important than the derivation, which is a bit tedious, so we'll begin with the answer and then, at the end of this section, show where it comes from.

First, we define the **specific heat ratio** γ (lowercase Greek gamma) to be

$$\gamma = \frac{C_P}{C_V} = \begin{cases} 1.67 & \text{monatomic gas} \\ 1.40 & \text{diatomic gas} \end{cases} \qquad (17.37)$$

The specific heat ratio has many uses in thermodynamics. Notice that γ is dimensionless.

An adiabatic process is one in which

$$pV^\gamma = \text{constant} \qquad \text{or} \qquad p_fV_f^\gamma = p_iV_i^\gamma \qquad (17.38)$$

This is similar to the isothermal $pV = \text{constant}$, but somewhat more complex due to the exponent γ.

The curves found by graphing $p = \text{constant}/V^\gamma$ are called **adiabats.** In **FIGURE 17.23** you see that the two adiabats are steeper than the hyperbolic isotherms. An adiabatic process moves along an adiabat in the same way that an isothermal process moves along an isotherm. You can see that the temperature falls during an adiabatic expansion and rises during an adiabatic compression.

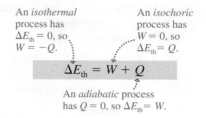

FIGURE 17.22 The relationship of three important processes to the first law of thermodynamics.

An *isothermal* process has $\Delta E_{th} = 0$, so $W = -Q$.

An *isochoric* process has $W = 0$, so $\Delta E_{th} = Q$.

$$\Delta E_{th} = W + Q$$

An *adiabatic* process has $Q = 0$, so $\Delta E_{th} = W$.

Actıv ONLINE Physıcs   8.6, 8.11

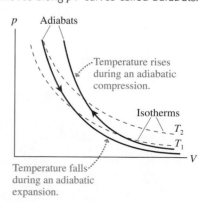

FIGURE 17.23 An adiabatic process moves along pV curves called *adiabats.*

Adiabats

Temperature rises during an adiabatic compression.

Isotherms

T_2

T_1

Temperature falls during an adiabatic expansion.

EXAMPLE 17.9 **An adiabatic compression**

Air containing gasoline vapor is admitted into the cylinder of an internal combustion engine at 1.00 atm pressure and 30°C. The piston rapidly compresses the gas from 500 cm³ to 50 cm³, a *compression ratio* of 10.

a. What are the final temperature and pressure of the gas?
b. Show the compression process on a *pV* diagram.
c. How much work is done to compress the gas?

MODEL The compression is rapid, with insufficient time for heat to be transferred from the gas to the environment, so we will model it as an adiabatic compression. We'll treat the gas as if it were 100% air.

SOLVE a. We know the initial pressure and volume, and we know the volume after the compression. For an adiabatic process, where pV^γ remains constant, the final pressure is

$$p_f = p_i \left(\frac{V_i}{V_f}\right)^\gamma = (1.00 \text{ atm})(10)^{1.40} = 25.1 \text{ atm}$$

Air is a mixture of N_2 and O_2, diatomic gases, so we used $\gamma = 1.40$. We can now find the temperature by using the ideal-gas law:

$$T_f = T_i \frac{p_f}{p_i} \frac{V_f}{V_i} = (303 \text{ K})(25.1)\left(\frac{1}{10}\right) = 761 \text{ K} = 488°C$$

Temperature *must* be in kelvins for doing gas calculations such as these.

b. **FIGURE 17.24** shows the *pV* diagram. The 30°C and 488°C isotherms are included to show how the temperature changes during the process.

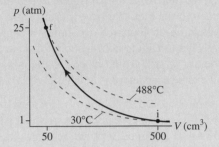

FIGURE 17.24 The adiabatic compression of the gas in an internal combustion engine.

c. The work done is $W = nC_V\Delta T$, with $\Delta T = 458$ K. The number of moles is found from the ideal-gas law and the initial conditions:

$$n = \frac{p_i V_i}{RT_i} = 0.0201 \text{ mol}$$

Thus the work done to compress the gas is

$$W = nC_V\Delta T = (0.0201 \text{ mol})(20.8 \text{ J/mol K})(458 \text{ K}) = 192 \text{ J}$$

ASSESS The temperature rises dramatically during the compression stroke of an engine. But the higher temperature has nothing to do with heat! **The temperature and thermal energy of the gas are increased not by heating the gas but by doing work on it.** This is an important idea to understand.

If we use the ideal-gas-law expression $p = nRT/V$ in the adiabatic equation $pV^\gamma = $ constant, we see that $TV^{\gamma-1}$ is also constant during an adiabatic process. Thus another useful equation for adiabatic processes is

$$T_f V_f^{\gamma-1} = T_i V_i^{\gamma-1} \tag{17.39}$$

Proof of Equation 17.38

Now let's see where Equation 17.38 comes from. Consider an adiabatic process in which an infinitesimal amount of work dW done on a gas causes an infinitesimal change in the thermal energy. For an adiabatic process, with $dQ = 0$, the first law of thermodynamics is

$$dE_{th} = dW \tag{17.40}$$

We can use Equation 17.35, which is valid for *any* gas process, to write $dE_{th} = nC_V dT$. Earlier in the chapter we found that the work done during a small volume change is $dW = -p\,dV$. With these substitutions, Equation 17.40 becomes

$$nC_V dT = -p\,dV \tag{17.41}$$

The ideal-gas law can now be used to write $p = nRT/V$. The n cancels, and the C_V can be moved to the other side of the equation to give

$$\frac{dT}{T} = -\frac{R}{C_V}\frac{dV}{V} \tag{17.42}$$

We're going to integrate Equation 17.42, but anticipating the need for $\gamma = C_P/C_V$ we can first use the fact that $C_P = C_V + R$ to write

$$\frac{R}{C_V} = \frac{C_P - C_V}{C_V} = \frac{C_P}{C_V} - 1 = \gamma - 1 \qquad (17.43)$$

Now we integrate Equation 17.42 from the initial state i to the final state f:

$$\int_{T_i}^{T_f} \frac{dT}{T} = -(\gamma - 1)\int_{V_i}^{V_f} \frac{dV}{V} \qquad (17.44)$$

Carrying out the integration gives

$$\ln\left(\frac{T_f}{T_i}\right) = \ln\left(\frac{V_i}{V_f}\right)^{\gamma-1} \qquad (17.45)$$

where we used the logarithm properties $\log a - \log b = \log(a/b)$ and $c \log a = \log(a^c)$.

Taking the exponential of both sides now gives

$$\left(\frac{T_f}{T_i}\right) = \left(\frac{V_i}{V_f}\right)^{\gamma-1} \qquad (17.46)$$

$$T_f V_f^{\gamma-1} = T_i V_i^{\gamma-1}$$

This was Equation 17.39. Writing $T = pV/nR$ and canceling $1/nR$ from both sides of the equation give Equation 17.38:

$$p_f V_f^{\gamma} = p_i V_i^{\gamma} \qquad (17.47)$$

This was a lengthy derivation, but it is good practice at seeing how the ideal-gas law and the first law of thermodynamics can work together to yield results of great importance.

STOP TO THINK 17.6   For the two processes shown, which of the following is true:

a. $Q_A > Q_B$
b. $Q_A = Q_B$
c. $Q_A < Q_B$

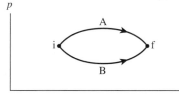

17.8 Heat-Transfer Mechanisms

You feel warmer when the sun is shining on you, colder when sitting on a metal bench or when the wind is blowing, especially if your skin is wet. This is due to the transfer of heat. Although we've talked about heat a lot in this chapter, we haven't said much about *how* heat is transferred from a hotter object to a colder object. There are four basic mechanisms by which objects exchange heat with their surroundings. Evaporation was treated in an earlier section; in this section, we will consider the other mechanisms.

Heat-transfer mechanisms

When two objects are in direct contact, such as the soldering iron and the circuit board, heat is transferred by *conduction*.

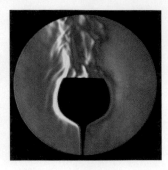

Air currents near a warm glass of water rise, taking thermal energy with them in a process known as *convection*.

The lamp at the top shines on the lambs huddled below, warming them. The energy is transferred by *radiation*.

Blowing on a hot cup of tea or coffee cools it by *evaporation*.

FIGURE 17.25 Conduction of heat through a solid.

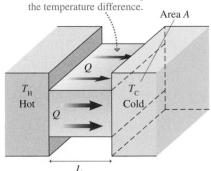

This material is conducting heat across the temperature difference.

Area A

Q

T_H Hot

T_C Cold

Q

L

TABLE 17.5 Thermal conductivities

Material	k (W/m K)
Diamond	2000
Silver	430
Copper	400
Aluminum	240
Iron	80
Stainless steel	14
Ice	1.7
Concrete	0.8
Glass	0.8
Styrofoam	0.035
Air (20°C, 1 atm)	0.023

Conduction

FIGURE 17.25 shows an object sandwiched between a higher temperature T_H and a lower temperature T_C. It makes no difference whether the object is wide and thin, such as a sheet of window glass separating a warm room from the cold outdoors, or long and skinny, such as a rod held in a flame. The temperature *difference* causes thermal energy to be transferred from the hot side to the cold side in a process known as **conduction.**

It is not surprising that more heat is transferred if the temperature difference ΔT is larger. A material with a larger cross section A (a fatter pipe) transfers more heat, while a thicker material, increasing the distance L between the hot and cold sources, decreases the rate of heat transfer.

These observations about heat conduction can be summarized in a single formula. If heat Q is transferred in a time interval Δt, the *rate* of heat transfer is $Q/\Delta t$. For a material of cross-section area A and length L, spanning a temperature difference $\Delta T = T_H - T_C$, the rate of heat transfer is

$$\frac{Q}{\Delta t} = k\frac{A}{L}\Delta T \qquad (17.48)$$

The quantity k, which characterizes whether the material is a good conductor of heat or a poor conductor, is called the **thermal conductivity** of the material. Because the heat-transfer rate J/s is a *power,* measured in watts, the units of k are W/m K. Values of k for common materials are given in Table 17.5; a material with a larger value of k is a better conductor of heat.

Most good heat conductors are metals, which are also good conductors of electricity. One exception is diamond. Although diamond is a poor electrical conductor, the strong bonds among atoms that make diamond such a hard material lead to a rapid transfer of thermal energy. Integrated circuits are often kept cool by by bonding them to metal (or sometimes diamond!) "heat sinks" that rapidly dissipate excess heat to the environment. Air and other gases are poor conductors of heat because there are no bonds between adjacent molecules.

EXAMPLE 17.10 Keeping a freezer cold

A 1.8-m-wide by 1.0-m-tall by 0.65-m-deep home freezer is insulated with 5.0-cm-thick Styrofoam insulation. At what rate must the compressor remove heat from the freezer to keep the inside at $-20°C$ in a room where the air temperature is $25°C$?

MODEL Heat is transferred through each of the six sides by conduction. The compressor must remove heat at the same rate it enters to maintain a steady temperature inside. The heat conduction is determined primarily by the thick insulation, so we'll neglect the thin inner and outer panels.

SOLVE Each of the six sides is a slab of Styrofoam with cross-section area A_i and thickness $L = 5.0$ cm. The total rate of heat transfer is

$$\frac{Q}{\Delta t} = \sum_{i=1}^{6} k\frac{A_i}{L}\Delta T = \frac{k\Delta T}{L}\sum_{i=1}^{6}A_i = \frac{k\Delta T}{L}A_{total}$$

The total surface area is

$$A_{total} = 2 \times (1.8 \text{ m} \times 1.0 \text{ m} + 1.8 \text{ m} \times 0.65 \text{ m}$$
$$+ 1.0 \text{ m} \times 0.65 \text{ m}) = 7.24 \text{ m}^2$$

Using $k = 0.035$ W/m K from Table 17.5, we find

$$\frac{Q}{\Delta t} = \frac{k\Delta t}{L}A_{total} = \frac{(0.035 \text{ W/m K})(45 \text{ K})(7.24 \text{ m}^2)}{0.050 \text{ m}} = 230 \text{ W}$$

Heat enters the freezer through the walls at the rate 230 J/s; thus the compressor must remove 230 J of heat energy every second to keep the temperature at $-20°C$.

ASSESS We'll learn in Chapter 19 how the compressor does this and how much work it must do. A typical freezer uses electric energy at a rate of about 150 W, so our result seems reasonable.

Thermal conductivity determines the *rate* at which heat energy is transferred. A metal chair *feels* colder to your bare skin than a wood chair, but is it? Both the metal and wood are at room temperature, but the metal has a much larger thermal conductivity and thus conducts heat out of your skin at a much higher rate. Your sensation of heat or cold is more closely connected with the rate of energy transfer than with the actual temperature.

Convection

Air is a poor conductor of heat, but thermal energy is easily transferred through air, water, and other fluids because the air and water can flow. A pan of water on the stove is heated at the bottom. This heated water expands, becomes less dense than the water above it, and thus rises to the surface, while cooler, denser water sinks to take its place. The same thing happens to air. This transfer of thermal energy by the motion of a fluid—the well-known idea that "heat rises"—is called **convection.**

Convection is usually the main mechanism for heat transfer in fluid systems. On a small scale, convection mixes the pan of water that you heat on the stove; on a large scale, convection is responsible for making the wind blow and ocean currents circulate. Air is a very poor thermal conductor, but it is very effective at transferring energy by convection. To use air for thermal insulation, it is necessary to trap the air in small pockets to limit convection. And that's exactly what feathers, fur, double-paned windows, and fiberglass insulation do. Convection is much more rapid in water than in air, which is why people can die of hypothermia in 68°F (20°C) water but can live quite happily in 68°F air.

Warm water (colored) moves by convection.

Radiation

The sun *radiates* energy to earth through the vacuum of space. Similarly, you feel the warmth from the glowing red coals in a fireplace.

All objects emit energy in the form of **radiation,** electromagnetic waves generated by oscillating electric charges in the atoms that form the object. These waves transfer energy from the object that emits the radiation to the object that absorbs it. Electromagnetic waves carry energy from the sun; this energy is absorbed when sunlight falls on your skin, warming you by increasing your thermal energy. Your skin also emits electromagnetic radiation, helping to keep your body cool by decreasing your thermal energy. Radiation is a significant part of the *energy balance* that keeps your body at the proper temperature.

This satellite image shows radiation emitted by the ocean waters off the east coast of the United States. You can clearly see the warm waters of the Gulf Stream, a large-scale convection that transfers heat to northern latitudes.

NOTE ▶ The word "radiation" comes from "radiate," meaning "to beam." Radiation can refer to x rays or to the radioactive decay of nuclei, but it also can refer simply to light and other forms of electromagnetic waves that "beam" from an object. Here we are using this second meaning of the term. ◀

You are familiar with radiation from objects hot enough to glow "red hot" or, at a high enough temperature, "white hot." The sun is a simply a very hot ball of glowing gas, and the white light from an incandescent lightbulb is radiation emitted by a thin wire filament heated to a very high temperature by an electric current. Objects at lower temperatures also radiate, but you can't see this radiation (although you can sometimes feel it) because it is long-wavelength infrared radiation.

Some films and detectors are infrared sensitive and can record the infrared radiation from objects. The false-color thermal image of a house that opened this chapter shows the infrared emission as the house radiates energy into the cooler environment. These images are used to assess where buildings need additional insulation.

The energy radiated by an object depends strongly on temperature. If heat energy Q is radiated in a time interval Δt by an object with surface area A and absolute temperature T, the *rate* of heat transfer is found to be

$$\frac{Q}{\Delta t} = e\sigma A T^4 \qquad (17.49)$$

Because the rate of energy transfer is power (1 J/s = 1 W), $Q/\Delta t$ is often called the *radiated power.* Notice the very strong fourth-power dependence on temperature. Doubling the absolute temperature of an object increases the radiated power by a factor of 16!

The parameter e in Equation 17.49 is the **emissivity** of the surface, a measure of how effectively it radiates. The value of e ranges from 0 to 1. σ is a constant, known as the Stefan-Boltzmann constant, with the value

$$\sigma = 5.67 \times 10^{-8} \text{ W/m}^2\text{ K}^4$$

NOTE ▶ Just as in the ideal-gas law, the temperature in Equation 17.49 *must* be in kelvins. ◀

Objects not only emit radiation, they also *absorb* radiation emitted by their surroundings. Suppose an object at temperature T is surrounded by an environment at temperature T_0. The *net* rate at which the object radiates heat energy—that is, radiation emitted minus radiation absorbed—is

$$\frac{Q_{\text{net}}}{\Delta t} = e\sigma A(T^4 - T_0^4) \qquad (17.50)$$

This makes sense. An object should have no *net* radiation if it's in thermal equilibrium ($T = T_0$) with its surroundings.

Notice that the emissivity e appears for absorption as well as emission; good emitters are also good absorbers. A perfect absorber ($e = 1$), one absorbing all light and radiation impinging on it but reflecting none, would appear completely black. Thus a perfect absorber is sometimes called a **black body.** But a perfect absorber would also be a perfect emitter, so thermal radiation from an ideal emitter is called **black-body radiation.** It seems strange that black objects are perfect emitters, but think of black charcoal glowing bright red in a fire. At room temperature, it "glows" equally bright with infrared.

EXAMPLE 17.11 Taking the sun's temperature

The radius of the sun is 6.96×10^8 m. At the distance of the earth, 1.50×10^{11} m, the intensity of solar radiation (measured by satellites above the atmosphere) is 1370 W/m². What is the temperature of the sun's surface?

MODEL Assume the sun to be an ideal radiator with $e = 1$.

SOLVE The total power radiated by the sun is the power per m² multiplied by the surface area of a sphere extending to the earth:

$$P = \frac{1370 \text{ W}}{1 \text{ m}^2} \times 4\pi (1.50 \times 10^{11} \text{ m})^2 = 3.87 \times 10^{26} \text{ W}$$

That is, the sun radiates energy at the rate $Q/\Delta t = 3.87 \times 10^{26}$ J/s. That's a lot of power! This energy is radiated from the surface of a sphere of radius R_S. Using this information in Equation 17.49, we find that the sun's surface temperature is

$$T = \left[\frac{Q/\Delta t}{e\sigma (4\pi R_S^2)} \right]^{1/4}$$

$$= \left[\frac{3.87 \times 10^{26} \text{ W}}{(1)(5.67 \times 10^{-8} \text{ W/m}^2 \text{ K}^4) 4\pi (6.96 \times 10^8 \text{ m})^2} \right]^{1/4}$$

$$= 5790 \text{ K}$$

ASSESS This temperature is confirmed by measurements of the solar spectrum, a topic we'll explore in Part VII.

Thermal radiation plays a prominent role in climate and global warming. The earth as a whole is in thermal equilibrium. Consequently, it must radiate back into space exactly as much energy as it receives from the sun. The incoming radiation from the hot sun is mostly visible light. The earth's atmosphere is transparent to visible light, so this radiation reaches the surface and is absorbed. The cooler earth radiates infrared radiation, but the atmosphere is *not* completely transparent to infrared. Some components of the atmosphere, notably water vapor and carbon dioxide, are strong absorbers of infrared radiation. They hinder the emission of radiation and, rather like a blanket, keep the earth's surface warmer than it would be without these gases in the atmosphere.

The **greenhouse effect,** as it's called, is a natural part of the earth's climate. The earth would be much colder and mostly frozen were it not for naturally occurring carbon dioxide in the atmosphere. But carbon dioxide also results from the burning of fossil fuels, and human activities since the beginning of the industrial revolution have increased the atmospheric concentration of carbon dioxide by nearly 50%. This human contribution has amplified the greenhouse effect and is the primary cause of global warming.

STOP TO THINK 17.7 Suppose you are an astronaut in space, hard at work in your sealed spacesuit. The only way that you can transfer excess heat to the environment is by

a. Conduction.   b. Convection.   c. Radiation.   d. Evaporation.

SUMMARY

The goals of Chapter 17 have been to expand our understanding of energy and to develop the first law of thermodynamics as a general statement of energy conservation.

General Principles

First Law of Thermodynamics

$$\Delta E_{th} = W + Q$$

The first law is a general statement of energy conservation.

Work W and heat Q depend on the process by which the system is changed.

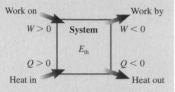

The change in the system depends only on the total energy exchanged $W + Q$, not on the process.

Energy

Thermal energy E_{th} Microscopic energy of moving molecules and stretched molecular bonds. ΔE_{th} depends on the initial/final states but is independent of the process.

Work W Energy transferred to the system by forces in a mechanical interaction.

Heat Q Energy transferred to the system via atomic-level collisions when there is a temperature difference. A thermal interaction.

Important Concepts

The **work** done on a gas is

$$W = -\int_{V_i}^{V_f} p \, dV$$

$$= -(\text{area under the } pV \text{ curve})$$

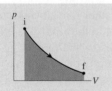

An **adiabatic process** is one for which $Q = 0$. Gases move along an **adiabat** for which $pV^\gamma = $ constant, where $\gamma = C_P/C_V$ is the **specific heat ratio.** An adiabatic process changes the temperature of the gas without heating it.

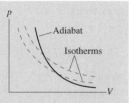

Calorimetry When two or more systems interact thermally, they come to a common final temperature determined by

$$Q_{net} = Q_1 + Q_2 + \cdots = 0$$

The **heat of transformation** L is the energy needed to cause 1 kg of substance to undergo a phase change

$$Q = \pm ML$$

The **specific heat** c of a substance is the energy needed to raise the temperature of 1 kg by 1 K:

$$Q = Mc\Delta T$$

The **molar specific heat** C is the energy needed to raise the temperature of 1 mol by 1 K:

$$Q = nC\Delta T$$

The molar specific heat of gases depends on the *process* by which the temperature is changed:

C_V = molar specific heat at **constant volume**
$C_P = C_V + R$ = molar specific heat at **constant pressure**

Heat is transferred by **conduction, convection, radiation,** and **evaporation.**

Conduction: $Q/\Delta t = (kA/L)\Delta T$

Radiation: $Q/\Delta t = e\sigma A T^4$

Summary of Basic Gas Processes

Process	Definition	Stays constant	Work	Heat
Isochoric	$\Delta V = 0$	V and p/T	$W = 0$	$Q = nC_V\Delta T$
Isobaric	$\Delta p = 0$	p and V/T	$W = -p\Delta V$	$Q = nC_P\Delta T$
Isothermal	$\Delta T = 0$	T and pV	$W = -nRT\ln(V_f/V_i)$	$\Delta E_{th} = 0$
Adiabatic	$Q = 0$	pV^γ	$W = \Delta E_{th}$	$Q = 0$
All gas processes	First law $\Delta E_{th} = W + Q = nC_V\Delta T$		Ideal-gas law $pV = nRT$	

Terms and Notation

internal energy, E_{int}	thermodynamic energy model	molar specific heat at constant	thermal conductivity, k
work, W	adiabatic process	volume, C_V	convection
mechanical interaction	specific heat, c	molar specific heat at constant	radiation
mechanical equilibrium	molar specific heat, C	pressure, C_P	emissivity, e
heat, Q	heat of transformation, L	specific heat ratio, γ	black body
thermal interaction	heat of fusion, L_f	adiabat	black-body radiation
thermal equilibrium	heat of vaporization, L_v	conduction	greenhouse effect
first law of thermodynamics	calorimetry		

 For homework assigned on MasteringPhysics, go to www.masteringphysics.com

Problem difficulty is labeled as | (straightforward) to ||| (challenging).

CONCEPTUAL QUESTIONS

1. When the space shuttle returns to earth, its surfaces get very hot as it passes through the atmosphere at high speed. Has the space shuttle been heated? If so, what was the source of the heat? If not, why is it hot?

2. Do (a) temperature, (b) heat, and (c) thermal energy describe a property of a system, an interaction of the system with its environment, or both? Explain.

3. The text says that the first law of thermodynamics is simply a general statement of the idea of conservation of energy. What does this mean? How does the first law embody the idea of energy conservation?

4. You have 100 g cubes labeled A and B. The cubes have equal densities and equal volumes, but A has a larger specific heat than B. Suppose cube A, initially at 0°C, is placed in good thermal contact with cube B, initially at 200°C, inside a well-insulated container. Is their final temperature greater than, less than, or equal to 100°C? Explain.

5. Two containers hold equal masses of nitrogen gas at equal temperatures. You supply 10 J of heat to container A while not allowing its volume to change, and you supply 10 J of heat to container B while not allowing its pressure to change. After-ward, is temperature T_A greater than, less than, or equal to T_B? Explain.

6. You need to raise the temperature of a gas by 10°C. To use the least amount of heat energy, should you heat the gas at constant pressure or at constant volume? Explain.

7. *Why* is the molar specific heat of a gas at constant pressure larger than the molar specific heat at constant volume?

8. **FIGURE Q17.8** shows an adiabatic process.
   a. Is the final temperature higher than, lower than, or equal to the initial temperature?
   b. Is any heat energy added to or removed from the system in this process? Explain.

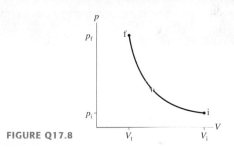

FIGURE Q17.8

9. **FIGURE Q17.9** shows two different processes taking an ideal gas from state i to state f. Is the work done on the gas in process A greater than, less than, or equal to the work done in process B? Explain.

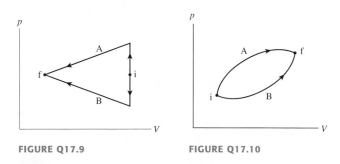

FIGURE Q17.9   **FIGURE Q17.10**

10. **FIGURE Q17.10** shows two different processes taking an ideal gas from state i to state f.
    a. Is the temperature *change* ΔT during process A larger than, smaller than, or equal to the change during process B? Explain.
    b. Is the heat energy added during process A greater than, less than, or equal to the heat added during process B? Explain.

11. Describe a series of steps in which you use the cylinder of Figure 17.13 to implement the ideal-gas process shown in **FIGURE Q17.11**. Then show the process as a first-law bar chart.

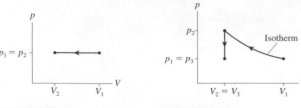

FIGURE Q17.11   **FIGURE Q17.12**

12. Describe a series of steps in which you use the cylinder of Figure 17.13 to implement the ideal-gas process shown in **FIGURE Q17.12**. Then show the process as a first-law bar chart.

13. The gas cylinder in **FIGURE Q17.13**, similar to the cylinder shown in Figure 17.13, is placed on a block of ice. The initial gas temperature is $> 0°C$.

   a. During the process that occurs until the gas reaches a new equilibrium, are (i) ΔT, (ii) W, and (iii) Q greater than, less than, or equal to zero? Explain.

   b. Draw a pV diagram showing the process.

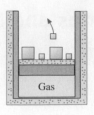

FIGURE Q17.13   **FIGURE Q17.14**

14. The gas cylinder in **FIGURE Q17.14** is similar to the cylinder described earlier in Figure 17.13, except that the bottom is insulated. Masses are slowly removed from the top of the piston until the total mass is reduced by 50%.

   a. During this process, are (i) ΔT (ii) W, and (iii) Q greater than, less than, or equal to zero? Explain.

   b. Draw a pV diagram showing the process.

EXERCISES AND PROBLEMS

Exercises

Section 17.1 It's All About Energy

Section 17.2 Work in Ideal-Gas Processes

1. | How much work is done on the gas in the process shown in **FIGURE EX17.1**?

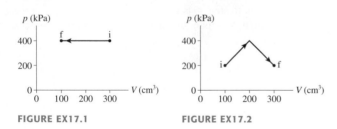

FIGURE EX17.1   **FIGURE EX17.2**

2. ‖ How much work is done on the gas in the process shown in **FIGURE EX17.2**?

3. ‖ 80 J of work are done on the gas in the process shown in **FIGURE EX17.3**. What is V_1 in cm³?

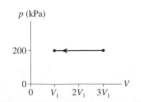

FIGURE EX17.3

4. ‖ A 2000 cm³ container holds 0.10 mol of helium gas at 300°C. How much work must be done to compress the gas to 1000 cm³ at (a) constant pressure and (b) constant temperature? (c) Show and label both processes on a single pV diagram.

Section 17.3 Heat

Section 17.4 The First Law of Thermodynamics

5. | Draw a first-law bar chart (see Figure 17.14) for the gas process in **FIGURE EX17.5**.

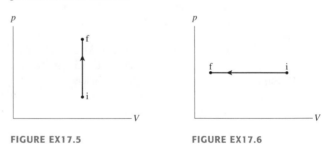

FIGURE EX17.5   **FIGURE EX17.6**

6. | Draw a first-law bar chart (see Figure 17.14) for the gas process in **FIGURE EX17.6**.

7. | Draw a first-law bar chart (see Figure 17.14) for the gas process in **FIGURE EX17.7**.

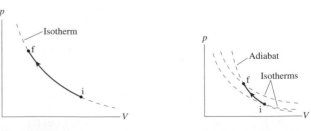

FIGURE EX17.7   **FIGURE EX17.8**

8. | Draw a first-law bar chart (see Figure 17.14) for the gas process in **FIGURE EX17.8**.

9. | 500 J of work are done on a system in a process that decreases the system's thermal energy by 200 J. How much heat energy is transferred to or from the system?

10. || A gas is compressed from 600 cm^3 to 200 cm^3 at a constant pressure of 400 kPa. At the same time, 100 J of heat energy is transferred out of the gas. What is the change in thermal energy of the gas during this process?

Section 17.5 Thermal Properties of Matter

11. || How much energy must be removed from a 6.0 cm × 6.0 cm × 6.0 cm block of ice to cool it from 0°C to −30°C?

12. | A rapidly spinning paddle wheel raises the temperature of 200 mL of water from 21°C to 25°C. How much (a) work is done and (b) heat is transferred in this process?

13. || How much heat is needed to change 20 g of mercury at 20°C into mercury vapor at the boiling point?

14. | a. 100 J of heat energy are transferred to 20 g of mercury. By how much does the temperature increase?
       b. How much heat is needed to raise the temperature of 20 g of water by the same amount?

15. || A beaker contains 200 mL of ethyl alcohol at 20°C. What is the minimum amount of energy that must be removed to produce solid ethyl alcohol?

16. || What is the maximum mass of lead you could melt with 1000 J of heat, starting from 20°C?

Section 17.6 Calorimetry

17. || 30 g of copper pellets are removed from a 300°C oven and immediately dropped into 100 mL of water at 20°C in an insulated cup. What will the new water temperature be?

18. | A copper block is removed from a 300°C oven and dropped into 1.00 L of water at 20.0°C. The water quickly reaches 25.5°C and then remains at that temperature. What is the mass of the copper block?

19. || A 50.0 g thermometer is used to measure the temperature of 200 mL of water. The specific heat of the thermometer, which is mostly glass, is 750 J/kg K, and it reads 20.0°C while lying on the table. After being completely immersed in the water, the thermometer's reading stabilizes at 71.2°C. What was the actual water temperature before it was measured?

20. || A 750 g aluminum pan is removed from the stove and plunged into a sink filled with 10.0 L of water at 20.0°C. The water temperature quickly rises to 24.0°C. What was the initial temperature of the pan in °C and in °F?

21. || A 500 g metal sphere is heated to 300°C, then dropped into a beaker containing 300 cm^3 of mercury at 20.0°C. A short time later the mercury temperature stabilizes at 99.0°C. Identify the metal.

Section 17.7 The Specific Heats of Gases

22. | A container holds 1.0 g of argon at a pressure of 8.0 atm.
     a. How much heat is required to increase the temperature by 100°C at constant volume?
     b. How much will the temperature increase if this amount of heat energy is transferred to the gas at constant pressure?

23. | A container holds 1.0 g of oxygen at a pressure of 8.0 atm.
     a. How much heat is required to increase the temperature by 100°C at constant pressure?
     b. How much will the temperature increase if this amount of heat energy is transferred to the gas at constant volume?

24. || The temperature of 2.0 g of helium is increased at constant volume by ΔT. What mass of oxygen can have its temperature increased by the same amount at constant volume using the same amount of heat?

25. | The volume of a gas is halved during an adiabatic compression that increases the pressure by a factor of 2.5.
     a. What is the specific heat ratio γ?
     b. By what factor does the temperature increase?

26. || A gas cylinder holds 0.10 mol of O_2 at 150°C and a pressure of 3.0 atm. The gas expands adiabatically until the pressure is halved. What are the final (a) volume and (b) temperature?

27. || A gas cylinder holds 0.10 mol of O_2 at 150°C and a pressure of 3.0 atm. The gas expands adiabatically until the volume is doubled. What are the final (a) pressure and (b) temperature?

Section 17.8 Heat-Transfer Mechanisms

28. | A 10 m × 14 m house is built on a 12-cm-thick concrete slab. What is the heat-loss rate through the slab if the ground temperature is 5°C while the interior of the house is 22°C?

29. | The ends of a 20-cm-long, 2.0-cm-diameter rod are maintained at 0°C and 100°C by immersion in an ice-water bath and boiling water. Heat is conducted through the rod at 4.5×10^4 J per hour. Of what material is the rod made?

30. || What maximum power can be radiated by a 10-cm-diameter solid lead sphere? Assume an emissivity of 1.

31. || Radiation from the head is a major source of heat loss from the human body. Model a head as a 20-cm-diameter, 20-cm-tall cylinder with a flat top. If the body's surface temperature is 35°C, what is the net rate of heat loss on a chilly 5°C day? All skin, regardless of color, is effectively black in the infrared where the radiation occurs, so use an emissivity of 0.95.

Problems

32. || A 5.0 g ice cube at −20°C is in a rigid, sealed container from which all the air has been evacuated. How much heat is required to change this ice cube into steam at 200°C?

33. || A 5.0-m-diameter garden pond is 30 cm deep. Solar energy is incident on the pond at an average rate of 400 W/m^2. If the water absorbs all the solar energy and does not exchange energy with its surroundings, how many hours will it take to warm from 15°C to 25°C?

34. || An 11 kg bowling ball at 0°C is dropped into a tub containing a mixture of ice and water. A short time later, when a new equilibrium has been established, there are 5.0 g less ice. From what height was the ball dropped? Assume no water or ice splashes out.

35. || The burner on an electric stove has a power output of 2.0 kW. A 750 g stainless steel teakettle is filled with 20°C water and placed on the already hot burner. If it takes 3.0 min for the water to reach a boil, what volume of water, in cm^3, was in the kettle? Stainless steel is mostly iron, so you can assume its specific heat is that of iron.

36. || Two cars collide head-on while each is traveling at 80 km/hr. Suppose all their kinetic energy is transformed into the thermal energy of the wrecks. What is the temperature increase of each car? You can assume that each car's specific heat is that of iron.

37. ||| 10 g of aluminum at 200°C and 20 g of copper are dropped into 50 cm^3 of ethyl alcohol at 15°C. The temperature quickly comes to 25°C. What was the initial temperature of the copper?

38. ‖ A 100 g ice cube at −10°C is placed in an aluminum cup whose initial temperature is 70°C. The system comes to an equilibrium temperature of 20°C. What is the mass of the cup?

39. ‖ 512 g of an unknown metal at a temperature of 15°C is dropped into a 100 g aluminum container holding 325 g of water at 98°C. A short time later, the container of water and metal stabilizes at a new temperature of 78°C. Identify the metal.

40. ‖ An experiment finds that the specific heat at constant volume of a monatomic gas is 625 J/kg K. Identify the gas.

41. ‖‖ A 150 L (≈ 40 gal) electric hot-water tank has a 5.0 kW heater. How many minutes will it take to raise the water temperature from 65°F to 140°F?

42. | An experiment measures the temperature of a 500 g substance while steadily supplying heat to it. **FIGURE P17.42** shows the results of the experiment. What are the (a) specific heat of the solid phase, (b) specific heat of the liquid phase, (c) melting and boiling temperatures, and (d) heats of fusion and vaporization?

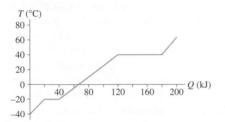

FIGURE P17.42

43. ‖ Liquid nitrogen is used in many low-temperature experiments. It is widely available, and cheaper than gasoline! How much heat must be removed from room temperature (20°C) nitrogen gas to produce 1.0 L of liquid nitrogen? The density of liquid nitrogen is 810 kg/m³.

44. ‖‖ Your 300 mL cup of coffee is too hot to drink when served at 90°C. What is the mass of an ice cube, taken from a −20°C freezer, that will cool your coffee to a pleasant 60°C?

45. ‖ You find an empty cooler, the kind used to keep drinks cold, at a Fourth of July picnic. The cooler has aluminum walls surrounded by insulating material. This is a 20 L cooler that uses 2.0 kg of aluminum. Just for fun, you toss in a firecracker, slam the lid, and sit on it to keep the lid from blowing off. A minute later, when you open the lid, you see that a built-in thermometer has risen from 25°C to 28°C. How much energy was released by the firecracker when it exploded?

46. | A typical nuclear reactor generates 1000 MW (1000 MJ/s) of electrical energy. In doing so, it produces 2000 MW of "waste heat" that must be removed from the reactor to keep it from melting down. Many reactors are sited next to large bodies of water so that they can use the water for cooling. Consider a reactor where the intake water is at 18°C. State regulations limit the temperature of the output water to 30°C so as not to harm aquatic organisms. How many liters of cooling water have to be pumped through the reactor each minute?

47. ‖ A beaker with a metal bottom is filled with 20 g of water at 20°C. It is brought into good thermal contact with a 4000 cm³ container holding 0.40 mol of a monatomic gas at 10 atm pressure. Both containers are well insulated from their surroundings. What is the gas pressure after a long time has elapsed?

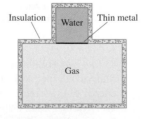

FIGURE P17.47

You can assume that the containers themselves are nearly massless and do not affect the outcome.

48. ‖ 2.0 mol of gas are at 30°C and a pressure of 1.5 atm. How much work must be done on the gas to compress it to one third of its initial volume at (a) constant temperature and (b) constant pressure? (c) Show both processes on a single *pV* diagram.

49. ‖ 500 J of work must be done to compress a gas to half its initial volume at constant temperature. How much work must be done to compress the gas by a factor of 10, starting from its initial volume?

50. ‖ A cylinder with a 16-cm-diameter piston contains gas at a pressure of 3.0 atm.
   a. How much force does the gas exert on the piston?
   b. How much force does the environment exert on the piston?
   c. The gas expands at constant pressure and pushes the piston out 10 cm. How much work is done by the environment?
   d. How much work is done by the gas?
   e. The thermal energy of the gas increases by 196 J in the expansion of part c. Was heat energy transferred to or from the gas in this process? How much?

51. ‖ A 10-cm-diameter cylinder contains argon gas at 10 atm pressure and a temperature of 50°C. A piston can slide in and out of the cylinder. The cylinder's initial length is 20 cm. 2500 J of heat are transferred to the gas, causing the gas to expand at constant pressure. What are (a) the final temperature and (b) the final length of the cylinder?

52. ‖ A cube 20 cm on each side contains 3.0 g of helium at 20°C. 1000 J of heat energy are transferred to this gas. What are (a) the final pressure if the process is at constant volume and (b) the final volume if the process is at constant pressure? (c) Show and label both processes on a single *pV* diagram.

53. ‖ An 8.0-cm-diameter, well-insulated vertical cylinder containing nitrogen gas is sealed at the top by a 5.1 kg frictionless piston. The air pressure above the piston is 100 kPa.
   a. What is the gas pressure inside the cylinder?
   b. Initially, the piston height above the bottom of the cylinder is 26 cm. What will be the piston height if an additional 3.5 kg are placed on top of the piston?

54. ‖ *n* moles of an ideal gas at temperature T_1 and volume V_1 expand isothermally until the volume has doubled. In terms of *n*, T_1, and V_1, what are (a) the final temperature, (b) the work done on the gas, and (c) the heat energy transferred to the gas?

55. ‖ 5.0 g of nitrogen gas at 20°C and an initial pressure of 3.0 atm undergo an isobaric expansion until the volume has tripled.
   a. What are the gas volume and temperature after the expansion?
   b. How much heat energy is transferred to the gas to cause this expansion?

   The gas pressure is then decreased at constant volume until the original temperature is reached.
   c. What is the gas pressure after the decrease?
   d. What amount of heat energy is transferred from the gas as its pressure decreases?
   e. Show the total process on a *pV* diagram. Provide an appropriate scale on both axes.

56. ‖ **FIGURE P17.56** shows two processes that take a gas from state i to state f. Show that $Q_A - Q_B = p_i V_i$.

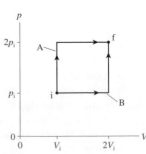

FIGURE P17.56

57. ‖ 0.10 mol of nitrogen gas follow the two processes shown in **FIGURE P17.57**. How much heat is required for each?

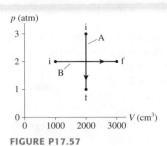

FIGURE P17.57

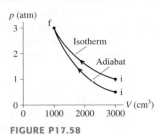

FIGURE P17.58

58. ‖ 0.10 mol of nitrogen gas follow the two processes shown in **FIGURE P17.58**. How much heat is required for each?

59. ‖ 0.10 mol of a monatomic gas follow the process shown in **FIGURE P17.59**.
   a. How much heat energy is transferred to or from the gas during process 1 → 2?
   b. How much heat energy is transferred to or from the gas during process 2 → 3?
   c. What is the total change in thermal energy of the gas?

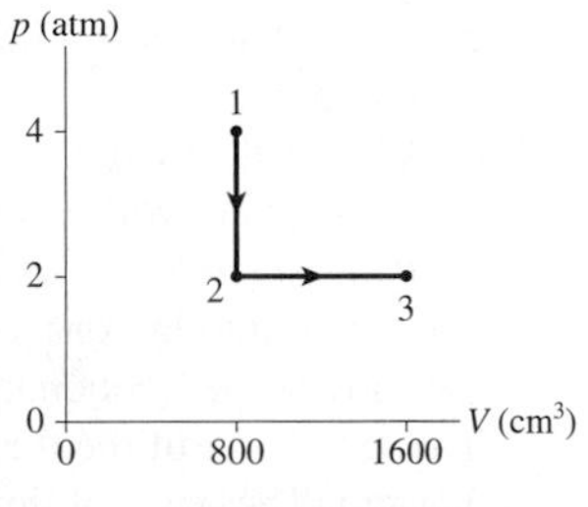

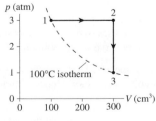

FIGURE P17.59

60. ‖ Two 800 cm³ containers hold identical amounts of a monatomic gas at 20°C. Container A is rigid. Container B has a 100 cm² piston with a mass of 10 kg that can slide up and down vertically without friction. Both containers are placed on identical heaters and heated for equal amounts of time.
   a. Will the final temperature of the gas in A be greater than, less than, or equal to the final temperature of the gas in B? Explain.
   b. Show both processes on a single *pV* diagram.
   c. What are the initial pressures in containers A and B?
   d. Suppose the heaters have 25 W of power and are turned on for 15 s. What is the final volume of container B?

61. ‖ Two cylinders each contain 0.10 mol of a diatomic gas at 300 K and a pressure of 3.0 atm. Cylinder A expands isothermally and cylinder B expands adiabatically until the pressure of each is 1.0 atm.
   a. What are the final temperature and volume of each?
   b. Show both processes on a single *pV* diagram. Use an appropriate scale on both axes.

62. ‖‖ A monatomic gas follows the process 1 → 2 → 3 shown in **FIGURE P17.62**. How much heat is needed for (a) process 1 → 2 and (b) process 2 → 3?

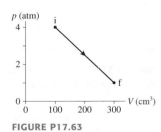

FIGURE P17.62

FIGURE P17.63

63. ‖ **FIGURE P17.63** shows a thermodynamic process followed by 0.015 mol of hydrogen.
   a. How much work is done on the gas?

b. By how much does the thermal energy of the gas change?
c. How much heat energy is transferred to the gas?

64. ‖ **FIGURE P17.64** shows a thermodynamic process followed by 120 mg of helium.
   a. Determine the pressure (in atm), temperature (in °C), and volume (in cm³) of the gas at points 1, 2, and 3. Put your results in a table for easy reading.
   b. How much work is done on the gas during each of the three segments?
   c. How much heat energy is transferred to or from the gas during each of the three segments?

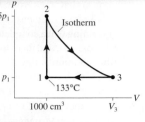

FIGURE P17.64

65. ‖ One cylinder in the diesel engine of a truck has an initial volume of 600 cm³. Air is admitted to the cylinder at 30°C and a pressure of 1.0 atm. The piston rod then does 400 J of work to rapidly compress the air. What are its final temperature and volume?

66. ‖ What compression ratios V_i/V_f will raise the temperature of (a) air and (b) argon from 30°C to 850°C in an adiabatic process?

67. ‖ a. What compression ratio V_{max}/V_{min} will raise the air temperature from 20°C to 1000°C in an adiabatic process?
   b. What pressure ratio p_{max}/p_{min} does this process have?

68. ‖ 2.0 g of helium at an initial temperature of 100°C and an initial pressure of 1.0 atm undergo an isobaric expansion until the volume has doubled. What are (a) the final temperature, (b) the work done on the gas, (c) the heat input to the gas, and (d) the change in thermal energy of the gas? (e) Show the process on a *pV* diagram, using proper scales on both axes.

69. ‖ 2.0 g of helium at an initial temperature of 100°C and an initial pressure of 1.0 atm undergo an isothermal expansion until the volume has doubled. What are (a) the final pressure, (b) the work done on the gas, (c) the heat input to the gas, and (d) the change in thermal energy of the gas? (e) Show the process on a *pV* diagram, using proper scales on both axes.

70. ‖ 14 g of nitrogen gas at STP are adiabatically compressed to a pressure of 20 atm. What are (a) the final temperature, (b) the work done on the gas, (c) the heat input to the gas, and (d) the compression ratio V_{max}/V_{min}? (e) Show the process on a *pV* diagram, using proper scales on both axes.

71. ‖ 14 g of nitrogen gas at STP are compressed in an isochoric process to a pressure of 20 atm. What are (a) the final temperature, (b) the work done on the gas, (c) the heat input to the gas, and (d) the pressure ratio p_{max}/p_{min}? (e) Show the process on a *pV* diagram, using proper scales on both axes.

72. ‖ When strong winds rapidly carry air down from mountains to a lower elevation, the air has no time to exchange heat with its surroundings. The air is compressed as the pressure rises, and its temperature can increase dramatically. These warm winds are called Chinook winds in the Rocky Mountains and Santa Ana winds in California. Suppose the air temperature high in the mountains behind Los Angeles is 0°C at an elevation where the air pressure is 60 kPa. What will the air temperature be, in °C and °F, when the Santa Ana winds have carried this air down to an elevation near sea level where the air pressure is 100 kPa?

73. ‖ You would like to put a solar hot water system on your roof, but you're not sure it's feasible. A reference book on solar energy shows that the ground-level solar intensity in your city is 800 W/m² for at least 5 hours a day throughout most of the year. Assuming that a completely black collector plate loses energy only by radiation, and that the air temperature is 20°C, what is the equilibrium temperature of a collector plate directly facing the sun? Note that while a plate has two sides, only the side facing the sun will radiate because the opposite side will be well insulated.

74. ‖ A cubical box 20 cm on a side is constructed from 1.2-cm-thick concrete panels. A 100 W lightbulb is sealed inside the box. What is the air temperature inside the box when the light is on if the surrounding air temperature is 20°C?

75. ‖ The sun's intensity at the distance of the earth is 1370 W/m². 30% of this energy is reflected by water and clouds; 70% is absorbed. What would be the earth's average temperature (in °C) if the earth had no atmosphere? The emissivity of the surface is very close to 1. (The actual average temperature of the earth, about 15°C, is higher than your calculation because of the greenhouse effect.)

In Problems 76 through 78 you are given the equation used to solve a problem. For each of these, you are to

a. Write a realistic problem for which this is the correct equation.
b. Finish the solution of the problem.

76. $50 \text{ J} = -n(8.31 \text{ J/mol K})(350 \text{ K})\ln\left(\dfrac{1}{3}\right)$

77. $(200 \times 10^{-6} \text{ m}^3)(13,600 \text{ kg/m}^3)$
$\qquad \times (140 \text{ J/kg K})(90°C - 15°C)$
$\qquad + (0.50 \text{ kg})(449 \text{ J/kg K})(90°C - T_i) = 0$

78. $(10 \text{ atm})V_2^{1.40} = (1.0 \text{ atm})V_1^{1.40}$

Challenge Problems

79. **FIGURE CP17.79** shows a thermodynamic process followed by 120 mg of helium.
   a. Determine the pressure (in atm), temperature (in °C), and volume (in cm³) of the gas at points 1, 2, and 3. Put your results in a table for easy reading.

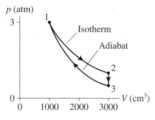

FIGURE CP17.79

b. How much work is done on the gas during each of the three segments?
c. How much heat is transferred to or from the gas during each of the three segments?

80. A 6.0-cm-diameter cylinder of nitrogen gas has a 4.0-cm-thick movable copper piston. The cylinder is oriented vertically, as shown in **FIGURE CP17.80**, and the air above the piston is evacuated. When the gas temperature is 20°C, the piston floats 20 cm above the bottom of the cylinder.
   a. What is the gas pressure?
   b. How many gas molecules are in the cylinder?

   Then 2.0 J of heat energy are transferred to the gas.
   c. What is the new equilibrium temperature of the gas?
   d. What is the final height of the piston?
   e. How much work is done on the gas as the piston rises?

FIGURE CP17.80

81. You come into lab one day and find a well-insulated 2000 mL thermos bottle containing 500 mL of boiling liquid nitrogen. The remainder of the thermos has nitrogen gas at a pressure of 1.0 atm. The gas and liquid are in thermal equilibrium. While waiting for lab to start, you notice a piece of iron on the table with "197 g" written on it. Just for fun, you drop the iron into the thermos and seal the cap tightly so that no gas can escape. After a few seconds have passed, what is the pressure inside the thermos? The density of liquid nitrogen is 810 kg/m³.

82. A cylindrical copper rod and an iron rod with exactly the same dimensions are welded together end to end. The outside end of the copper rod is held at 100°C, and the outside end of the iron rod is held at 0°C. What is the temperature at the midpoint where the rods are joined together?

83. 0.020 mol of a diatomic gas, with initial temperature 20°C, are compressed from 1500 cm³ to 500 cm³ in a process in which $pV^2 = $ constant.
   a. What is the final temperature (in °C)?
   b. How much heat is added during this process?
   c. Draw the pV diagram, include proper scales on both axes.

STOP TO THINK ANSWERS

Stop to Think 17.1: a. The piston does work W on the gas. There's no heat because of the insulation, and $\Delta E_{mech} = 0$ because the gas as a whole doesn't move. Thus $\Delta E_{th} = W > 0$. The work increases the system's thermal energy and thus raises its temperature.

Stop to Think 17.2: d. $W_A = 0$ because A is an isochoric process. $W_B = W_{1 \text{ to } 2} + W_{2 \text{ to } 3}$. $|W_{2 \text{ to } 3}| > |W_{1 \text{ to } 2}|$ because there's more area under the curve, and $W_{2 \text{ to } 3}$ is positive whereas $W_{1 \text{ to } 2}$ is negative. Thus W_B is positive.

Stop to Think 17.3: b and e. The temperature rises in d from doing work on the gas ($\Delta E_{th} = W$), not from heat. e involves heat because there is a temperature difference. The temperature of the gas doesn't change because the heat is used to do the work of lifting a weight.

Stop to Think 17.4: c. The temperature increases so E_{th} must increase. W is negative in an expansion, so Q must be positive and larger than $|W|$.

Stop to Think 17.5: a. A has a smaller specific heat and thus less thermal inertia. The temperature of A will change more than the temperature of B.

Stop to Think 17.6: a. $W_A + Q_A = W_B + Q_B$. The area under process A is larger than the area under B, so W_A is *more negative* than W_B. Q_A has to be more positive than Q_B to maintain the equality.

Stop to Think 17.7: c. Conduction, convection, and evaporation require matter. Only radiation transfers energy through the vacuum of space.

18 The Micro/Macro Connection

Heating the air in a hot-air balloon increases the thermal energy of the air molecules. This causes the gas to expand, lowering its density and allowing the balloon to float in the cooler surrounding air.

▶ **Looking Ahead**
The goal of Chapter 18 is to understand the properties of a macroscopic system in terms of the microscopic behavior of its molecules. In this chapter you will learn to:

- Understand how molecular motions and collisions are responsible for macroscopic phenomena such as pressure and heat transfer.
- Establish a connection among temperature, thermal energy, and the average translational kinetic energy of the molecules in the system.
- Use the micro/macro connection to predict the molar specific heats of gases and solids.
- Use the second law of thermodynamics to understand how interacting systems come to thermal equilibrium.

◀ **Looking Back**
The material in this chapter depends on understanding heat, thermal energy, and the properties of ideal gases. Please review:

- Sections 16.5–16.6 Ideal gases.
- Sections 17.3–17.4 Heat and the first law of thermodynamics.
- Sections 17.5 and 17.7 Specific heats and molar specific heats.

A gas consists of a vast number of molecules ceaselessly colliding with each other and the walls of their container as they whiz about. A solid contains uncountable atoms vibrating around their equilibrium positions. Our goal in this chapter is to show how this turmoil at the microscopic level gives rise to predictable and steady values of macroscopic variables such as pressure, temperature, and specific heat.

This micro/macro connection, which goes by the more formal name **kinetic theory,** will help us elucidate several puzzles that we noted in the previous two chapters. For example, why do all elemental solids have the same molar specific heats, as do all monatomic gases and all diatomic gases? Kinetic theory will also give us a better understanding of *heat* and of how it is that two systems come to thermal equilibrium as they interact.

We'll also introduce a new law of nature, the second law of thermodynamics. The second law is rather subtle, but it has profound implications. We'll use the second law to understand why it is that heat energy "flows" from hot to cold rather than from cold to hot.

18.1 Molecular Speeds and Collisions

Let us begin by thinking about gases at the atomic level. If gases really are composed of atoms and molecules in motion, how fast are the molecules moving? Do all molecules move with the same speed, or is there a range of speeds?

FIGURE 18.1 An experiment to measure the speeds of molecules in a gas.

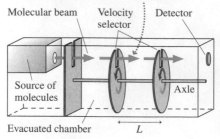

The only molecules that reach the detector are those whose speed allows them to travel distance L during the time it takes the disks to make one full revolution.

FIGURE 18.2 The distribution of molecular speeds in a sample of nitrogen gas.

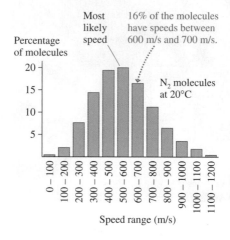

Speed range (m/s)

To answer this question, **FIGURE 18.1** shows an experiment to measure the speeds of molecules in a gas. The molecules emerging from the source form what is called a *molecular beam.* At the right end, a detector counts the number of molecules that make it through the apparatus each second. The experiment takes place inside a vacuum chamber, allowing the molecules to travel without undergoing any collisions.

The two rotating disks form a *velocity selector.* Once every revolution, the slot in the first disk allows a small pulse of molecules to pass through. By the time these molecules reach the second disk, the slots have rotated. The molecules can pass through the second slot and be detected *only* if they have exactly the right speed $v = L/\Delta t$ to travel the distance L between the two disks during time interval Δt it takes the axle to complete one revolution. Molecules having any other speed are blocked by the second disk and are not detected. By changing the rotation frequency of the axle, and thus changing Δt, this apparatus can measure how many molecules have each of many possible speeds.

FIGURE 18.2 shows the results for nitrogen gas (N_2) at $T = 20°C$. The data are presented in the form of a **histogram,** a bar chart in which the height of each bar tells how many (or, in this case, what percentage) of the molecules have a speed in the *range* of speeds shown below the bar. For example, 16% of the molecules have speeds in the range from 600 m/s to 700 m/s. All the bars sum to 100%, showing that this histogram describes *all* of the molecules leaving the source.

It turns out that the molecules have what is called a *distribution* of speeds, ranging from as low as ≈ 100 m/s to as high as ≈ 1200 m/s. But not all speeds are equally likely; there is a *most likely speed* of ≈ 550 m/s. This is really fast, ≈ 1200 mph! Notice that the majority of molecular speeds do not differ much from the most likely speed. Few molecules have very high or very low speeds, while well over 60% (sum of the center four bars) have speeds within the range 300 m/s to 700 m/s. Changing the temperature or changing to a different gas changes the most likely speed, as we'll learn later in the chapter, but it does not change the *shape* of the distribution.

If you were to repeat the experiment a few seconds or a few hours later, you would again find the most likely speed to be ≈ 550 m/s and that 16% of the molecules have speeds between 600 m/s and 700 m/s. Think about what this means. The "molecular deck of cards" is constantly being reshuffled by molecular collisions, causing some molecules to speed up and others to slow down, yet 16% of the molecules always have speeds between 600 m/s and 700 m/s.

There's an important lesson here. A gas consists of a vast number of molecules, each moving randomly and undergoing millions of collisions every second. Despite the apparent chaos, *averages,* such as the average number of molecules in the speed range 600 to 700 m/s, have precise, predictable values. **The micro/macro connection is built on the idea that the macroscopic properties of a system, such as temperature or pressure, are related to the *average* behavior of the atoms and molecules.**

Mean Free Path

Imagine someone opening a bottle of strong perfume a few feet away from you. If molecular speeds are hundreds of meters per second, you might expect to smell the perfume almost instantly. But that isn't what happens. As you know, it takes many seconds for the molecules to *diffuse* across the room. Let's see why this is.

FIGURE 18.3 shows a "movie" of one molecule as it moves through a gas. Instead of zipping along in a straight line, as it would in a vacuum, the molecule follows a convoluted zig-zag path in which it frequently collides with other molecules. A molecule may have traveled hundreds of meters by the time it manages to get 1 or 2 m away from its starting point.

The random distribution of the molecules in the gas causes the straight-line segments between collisions to be of unequal lengths. A question we could ask is: What is

FIGURE 18.3 A single molecule follows a zig-zag path through a gas as it collides with other molecules.

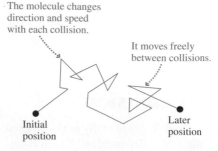

The molecule changes direction and speed with each collision.

It moves freely between collisions.

Initial position

Later position

the *average* distance between collisions? If a molecule has N_{coll} collisions as it travels distance L, the average distance between collisions, which is called the **mean free path** λ (lowercase Greek lambda), is

$$\lambda = \frac{L}{N_{coll}} \qquad (18.1)$$

The concept of mean free path is used not only in gases but also to describe electrons moving through conductors and light passing through a medium that scatters the photons.

Our task is to determine the number of collisions. FIGURE 18.4a shows two molecules approaching each other. We will assume that the molecules are spherical and of radius r. We will also continue the ideal-gas assumption that the molecules undergo hard-sphere collisions, like billiard balls. In that case, the molecules will collide if the distance between their *centers* is less than $2r$. They will miss if the distance is greater than $2r$.

FIGURE 18.4b shows a cylinder of radius $2r$ centered on the trajectory of a "sample" molecule. The sample molecule collides with any "target" molecule whose center is located within the cylinder, causing the cylinder to bend at that point. Hence the number of collisions N_{coll} is equal to the number of molecules in a cylindrical volume of length L.

The volume of a cylinder is $V_{cyl} = AL = \pi(2r)^2 L$. If the number density of the gas is N/V particles per m^3, then the number of collisions along a trajectory of length L is

$$N_{coll} = \frac{N}{V}V_{cyl} = \frac{N}{V}\pi(2r)^2L = 4\pi\frac{N}{V}r^2L \qquad (18.2)$$

Thus the mean free path between collisions is

$$\lambda = \frac{L}{N_{coll}} = \frac{1}{4\pi(N/V)r^2}$$

We made a tacit assumption in this derivation that the target molecules are at rest. While the general idea behind our analysis is correct, a more careful calculation in which all molecules move introduces an extra factor of $\sqrt{2}$, giving

$$\lambda = \frac{1}{4\sqrt{2}\pi(N/V)r^2} \qquad \text{(mean free path)} \qquad (18.3)$$

Measurements are necessary to determine precise values of atomic and molecular radii, but a reasonable rule of thumb is to assume that atoms in a monatomic gas have $r \approx 0.5 \times 10^{-10}$ m and diatomic molecules have $r \approx 1.0 \times 10^{-10}$ m.

FIGURE 18.4 A sample molecule will collide with all target molecules whose centers are within a bent cylinder of radius $2r$ centered on its path.

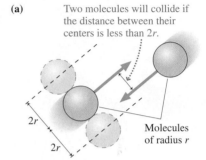

(a) Two molecules will collide if the distance between their centers is less than $2r$.

$2r$

$2r$

Molecules of radius r

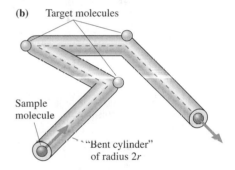

(b) Target molecules

Sample molecule

"Bent cylinder" of radius $2r$

EXAMPLE 18.1 The mean free path at room temperature

What is the mean free path of a nitrogen molecule at 1.0 atm pressure and room temperature (20°C)?

SOLVE Nitrogen is a diatomic molecule, so $r \approx 1.0 \times 10^{-10}$ m. We can use the ideal-gas law in the form $pV = Nk_BT$ to determine the number density:

$$\frac{N}{V} = \frac{p}{k_BT} = \frac{101{,}300 \text{ Pa}}{(1.38 \times 10^{-23} \text{ J/K})(293 \text{ K})} = 2.5 \times 10^{25} \text{ m}^{-3}$$

Thus the mean free path is

$$\lambda = \frac{1}{4\sqrt{2}\pi(N/V)r^2}$$

$$= \frac{1}{4\sqrt{2}\pi(2.5 \times 10^{25} \text{ m}^{-3})(1.0 \times 10^{-10} \text{ m})^2}$$

$$= 2.3 \times 10^{-7} \text{ m} = 230 \text{ nm}$$

ASSESS You learned in Example 16.5 that the average separation between gas molecules at STP is ≈ 5.7 nm. It seems that any given molecule can slip between its neighbors, which are spread out in three dimensions, and travel—on average—about 40 times the average spacing before it collides with another molecule.

STOP TO THINK 18.1 The table shows the properties of four gases, each having the same number of molecules. Rank in order, from largest to smallest, the mean free paths λ_A to λ_D of molecules in these gases.

Gas	A	B	C	D
Volume	V	$2V$	V	V
Atomic mass	m	m	$2m$	m
Atomic radius	r	r	r	$2r$

18.2 Pressure in a Gas

Why does a gas have pressure? In Chapter 15, where pressure was introduced, we suggested that the pressure in a gas is due to collisions of the molecules with the walls of its container. The force due to one such collision may be unmeasurably tiny, but the steady rain of a vast number of molecules striking a wall each second exerts a measurable macroscopic force. The gas pressure is the force per unit area ($p = F/A$) resulting from these molecular collisions.

Our task in this section is to calculate the pressure by doing the appropriate averaging over molecular motions and collisions. This task can be divided into three main pieces:

1. Calculate the impulse a single molecule exerts on the wall during a collision.
2. Find the force due to all collisions.
3. Introduce an appropriate average speed.

Force Due to a Single Collision

FIGURE 18.5 A molecule colliding with the wall exerts an impulse on it.

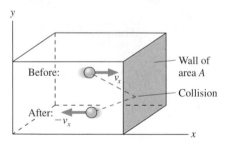

FIGURE 18.5 shows a molecule with an x-component of velocity v_x approaching a wall. We will assume that the collision with the wall is perfectly elastic, an assumption we will justify later, in which case the molecule rebounds from the wall with its x-component of velocity changed from $+v_x$ to $-v_x$. This molecule experiences an impulse. We can use the impulse-momentum theorem from Chapter 9 to write

$$(J_x)_{\text{wall on molecule}} = \Delta p = m(-v_x) - mv_x = -2mv_x \qquad (18.4)$$

According to Newton's third law, the wall experiences the equal but opposite impulse

$$(J_x)_{\text{molecule on wall}} = +2mv_x \qquad (18.5)$$

as a result of this single collision.

Suppose there are N_{coll} such collisions during a very small time interval Δt. If we assume for the moment that all molecules have the *same* x-component velocity v_x, the net impulse of these collisions on the wall is

$$J_{\text{wall}} = N_{\text{coll}} \times (J_x)_{\text{molecule on wall}} = 2N_{\text{coll}}mv_x \qquad (18.6)$$

FIGURE 18.6 Impulse is the area under the force-versus-time curve.

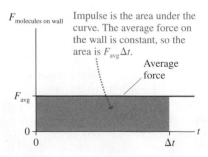

Impulse is the area under the curve. The average force on the wall is constant, so the area is $F_{\text{avg}}\Delta t$.

Average force

FIGURE 18.6 reminds you that impulse is the area under the force-versus-time curve and thus $J_{\text{wall}} = F_{\text{avg}}\Delta t$, where F_{avg} is the *average* force exerted on the wall. Using this in Equation 18.6, we see that the average force on the wall due to many molecular collisions is

$$F_{\text{avg}} = 2\frac{N_{\text{coll}}}{\Delta t}mv_x \qquad (18.7)$$

The quantity $N_{\text{coll}}/\Delta t$ is the *rate* of collisions with the wall—that is, the number of collisions per second. **FIGURE 18.7** shows how to determine the rate of collisions. Let the time interval Δt be much less than the average time between collisions, so no collisions alter the molecular speeds during this interval. (This assumption about Δt isn't really necessary, but it makes it easier to think about what's going on.) During Δt, all molecules travel distance $\Delta x = v_x \Delta t$ along the x-axis. This distance is shaded in the figure. *Every one* of the molecules in this shaded region that is moving to the right will reach and collide with the wall during time Δt. Molecules outside this region will not reach the wall during Δt and will not collide.

The shaded region has volume $A\Delta x$, where A is the surface area of the wall. Only half the molecules are moving to the right, hence the number of collisions during Δt is

$$N_{\text{coll}} = \frac{1}{2}\frac{N}{V}A\Delta x = \frac{1}{2}\frac{N}{V}Av_x\Delta t \qquad (18.8)$$

and thus the rate of collisions is

$$\frac{N_{\text{coll}}}{\Delta t} = \frac{1}{2}\frac{N}{V}Av_x \qquad (18.9)$$

The average force on the wall is found by substituting $N_{\text{coll}}/\Delta t$ from Equation 18.9 into Equation 18.7:

$$F_{\text{avg}} = 2\left(\frac{1}{2}\frac{N}{V}Av_x\right)mv_x = \frac{N}{V}mv_x^2 A \qquad (18.10)$$

Notice that this expression for F_{avg} does not depend on any details of the molecular collisions.

We can relax the assumption that all molecules have the same speed by replacing the squared velocity v_x^2 in Equation 18.10 with its average value. That is,

$$F_{\text{avg}} = \frac{N}{V}m(v_x^2)_{\text{avg}}A \qquad (18.11)$$

where $(v_x^2)_{\text{avg}}$ is the quantity v_x^2 averaged over all the molecules in the container.

The Root-Mean-Square Speed

We need to be somewhat careful when averaging velocities. The velocity component v_x has a sign. At any instant of time, half the molecules in a container move to the right and have positive v_x while the other half move to the left and have negative v_x. Thus the *average velocity* is $(v_x)_{\text{avg}} = 0$. If this weren't true, the entire container of gas would move away!

The speed of a molecule is $v = (v_x^2 + v_y^2 + v_z^2)^{1/2}$. Thus the average of the speed squared is

$$(v^2)_{\text{avg}} = (v_x^2 + v_y^2 + v_z^2)_{\text{avg}} = (v_x^2)_{\text{avg}} + (v_y^2)_{\text{avg}} + (v_z^2)_{\text{avg}} \qquad (18.12)$$

The square root of $(v^2)_{\text{avg}}$ is called the **root-mean-square speed** v_{rms}:

$$v_{\text{rms}} = \sqrt{(v^2)_{\text{avg}}} \qquad \text{(root-mean-square speed)} \qquad (18.13)$$

This is usually called the *rms speed*. You can remember its definition by noting that its name is the *opposite* of the sequence of operations: First you square all the speeds, then you average the squares (find the mean), then you take the square root. Because the square root "undoes" the square, v_{rms} must, in some sense, give an average speed.

> **NOTE** ▶ We could compute a true average speed v_{avg}, but that calculation is difficult. More important, the root-mean-square speed tends to arise naturally in many scientific and engineering calculations. It turns out that v_{rms} differs from v_{avg} by less than 10%, so for practical purposes we can interpret v_{rms} as being essentially the average speed of a molecule in a gas. ◀

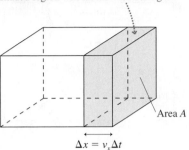

FIGURE 18.7 Determining the rate of collisions.

Only molecules moving to the right in the shaded region will hit the wall during Δt.

Area A

$\Delta x = v_x \Delta t$

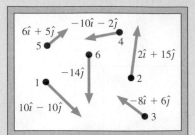

FIGURE 18.8 The molecular velocities of Example 18.2. Units are m/s.

EXAMPLE 18.2 **Calculating the root-mean-square speed**

FIGURE 18.8 shows the velocities of all the molecules in a six-molecule, two-dimensional gas. Calculate and compare the average velocity $\vec{v}_{avg}$, the average speed v_{avg}, and the rms speed v_{rms}.

SOLVE Table 18.1 shows the velocity components v_x and v_y for each molecule, the squares v_x^2 and v_y^2, their sum $v^2 = v_x^2 + v_y^2$, and the speed $v = (v_x^2 + v_y^2)^{1/2}$. Averages of all the values in each column are shown at the bottom. You can see that the average velocity is $\vec{v}_{avg} = \vec{0}$ m/s and the average speed is $v_{avg} = 11.9$ m/s. The rms speed is

$$v_{rms} = \sqrt{(v^2)_{avg}} = \sqrt{148.3 \text{ m}^2/\text{s}^2} = 12.2 \text{ m/s}$$

ASSESS The rms speed is only 2.5% greater than the average speed.

TABLE 18.1 Calculation of rms speed and average speed for the molecules of Example 18.2

Molecule	v_x	v_y	v_x^2	v_y^2	v^2	v
1	10	−10	100	100	200	14.1
2	2	15	4	225	229	15.1
3	−8	6	64	36	100	10.0
4	−10	−2	100	4	104	10.2
5	6	5	36	25	61	7.8
6	0	−14	0	196	196	14.0
Average	0	0			148.3	11.9

There's nothing special about the x-axis. The coordinate system is something that *we* impose on the problem, so *on average* it must be the case that

$$(v_x^2)_{avg} = (v_y^2)_{avg} = (v_z^2)_{avg} \qquad (18.14)$$

Hence we can use Equation 18.12 and the definition of v_{rms} to write

$$v_{rms}^2 = (v_x^2)_{avg} + (v_y^2)_{avg} + (v_z^2)_{avg} = 3(v_x^2)_{avg} \qquad (18.15)$$

Consequently, $(v_x^2)_{avg}$ is

$$(v_x^2)_{avg} = \frac{1}{3}v_{rms}^2 \qquad (18.16)$$

Using this result in Equation 18.11 gives us the net force on the wall of the container:

$$F_{net} = \frac{1}{3}\frac{N}{V}mv_{rms}^2 A \qquad (18.17)$$

Thus the pressure on the wall of the container due to all the molecular collisions is

$$p = \frac{F}{A} = \frac{1}{3}\frac{N}{V}mv_{rms}^2 \qquad (18.18)$$

We have met our goal. Equation 18.18 expresses the macroscopic pressure in terms of the microscopic physics. The pressure depends on the number density of molecules in the container and on how fast, on average, the molecules are moving.

EXAMPLE 18.3 The rms speed of helium atoms

A container holds helium at a pressure of 200 kPa and a temperature of 60.0°C. What is the rms speed of the helium atoms?

SOLVE The rms speed can be found from the pressure and the number density. Using the ideal-gas law gives us the number density:

$$\frac{N}{V} = \frac{p}{k_B T} = \frac{200{,}000 \text{ Pa}}{(1.38 \times 10^{-23} \text{ J/K})(333 \text{ K})} = 4.35 \times 10^{25} \text{ m}^{-3}$$

The mass of a helium atom is $m = 4 \text{ u} = 6.64 \times 10^{-27}$ kg. Thus

$$v_{rms} = \sqrt{\frac{3p}{(N/V)m}} = 1440 \text{ m/s}$$

STOP TO THINK 18.2 The speed of every molecule in a gas is suddenly increased by a factor of 4. As a result, v_{rms} increases by a factor of

a. 2.

b. <4 but not necessarily 2.

c. 4.

d. >4 but not necessarily 16.

e. 16.

f. v_{rms} doesn't change.

18.3 Temperature

A molecule of mass m and velocity v has translational kinetic energy

$$\epsilon = \frac{1}{2}mv^2 \tag{18.19}$$

We'll use ϵ (lowercase Greek epsilon) to distinguish the energy of a molecule from the system energy E. Thus the average translational kinetic energy is

$$\epsilon_{avg} = \text{average translational kinetic energy of a molecule}$$
$$= \frac{1}{2}m(v^2)_{avg} = \frac{1}{2}mv_{rms}^2 \tag{18.20}$$

We've included the word "translational" to distinguish ϵ from rotational kinetic energy, which we will consider later in this chapter.

We can write the gas pressure, Equation 18.18, in terms of the average translational kinetic energy as

$$p = \frac{2}{3}\frac{N}{V}\left(\frac{1}{2}mv_{rms}^2\right) = \frac{2}{3}\frac{N}{V}\epsilon_{avg} \tag{18.21}$$

The pressure is directly proportional to the average molecular translational kinetic energy. This makes sense. More-energetic molecules will hit the walls harder as they bounce and thus exert more force on the walls.

It's instructive to write Equation 18.21 as

$$pV = \frac{2}{3}N\epsilon_{avg} \tag{18.22}$$

We know, from the ideal-gas law, that

$$pV = Nk_B T \tag{18.23}$$

Comparing these two equations, we reach the significant conclusion that the average translational kinetic energy per molecule is

$$\epsilon_{avg} = \frac{3}{2}k_B T \qquad \text{(average translational kinetic energy)} \tag{18.24}$$

where the temperature T is in kelvins. For example, the average translational kinetic energy of a molecule at room temperature (20°C) is

$$\epsilon_{avg} = \frac{3}{2}(1.38 \times 10^{-23} \text{ J/K})(293 \text{ K}) = 6.1 \times 10^{-21} \text{ J}$$

NOTE ▶ A molecule's average translational kinetic energy depends *only* on the temperature, not on the molecule's mass. If two gases have the same temperature, their molecules have the same average translational kinetic energy. This will be an important idea when we look at the thermal interaction between two systems. ◀

Equation 18.24 is especially satisfying because it finally gives real meaning to the concept of temperature. Writing it as

$$T = \frac{2}{3k_B}\epsilon_{avg} \qquad (18.25)$$

we can see that, for a gas, **this thing we call *temperature* measures the average translational kinetic energy.** A higher temperature corresponds to a larger value of ϵ_{avg} and thus to higher molecular speeds. This concept of temperature also gives meaning to *absolute zero* as the temperature at which $\epsilon_{avg} = 0$ and all molecular motion ceases. (Quantum effects at very low temperatures prevent the motions from actually stopping, but our classical theory predicts that they would.) **FIGURE 18.9** summarizes what we've learned thus far about the micro/macro connection.

We can now justify our assumption that molecular collisions are perfectly elastic. Suppose they were not. That is, suppose that a small amount of kinetic energy was lost in each collision. If that were so, the average translational kinetic energy ϵ_{avg} of the gas would slowly decrease and we would see a steadily decreasing temperature. But that doesn't happen. The temperature of an isolated system remains perfectly constant, indicating that ϵ_{avg} is not changing with time. Consequently, the collisions must be perfectly elastic.

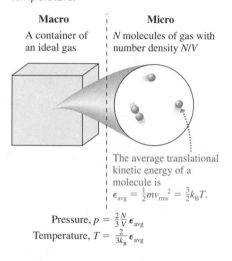

FIGURE 18.9 The micro/macro connection for pressure and temperature.

Macro

A container of an ideal gas

Micro

N molecules of gas with number density N/V

The average translational kinetic energy of a molecule is $\epsilon_{avg} = \frac{1}{2}mv_{rms}^2 = \frac{3}{2}k_BT$.

Pressure, $p = \frac{2}{3}\frac{N}{V}\epsilon_{avg}$

Temperature, $T = \frac{2}{3k_B}\epsilon_{avg}$

EXAMPLE 18.4 Total microscopic kinetic energy

What is the total translational kinetic energy of the molecules in 1.0 mol of gas at STP?

SOLVE The average translational kinetic energy of each molecule is

$$\epsilon_{avg} = \frac{3}{2}k_BT = \frac{3}{2}(1.38 \times 10^{-23} \text{ J/K})(273 \text{ K})$$

$$= 5.65 \times 10^{-21} \text{ J}$$

1.0 mol of gas contains N_A molecules; hence the total kinetic energy is

$$K_{micro} = N_A\epsilon_{avg} = 3400 \text{ J}$$

ASSESS The energy of any one molecule is incredibly small. Nonetheless, a macroscopic system has substantial thermal energy because it consists of an incredibly large number of molecules.

8.1–8.3 **Activ** Physics

By definition, $\epsilon_{avg} = \frac{1}{2}mv_{rms}^2$. Using the ideal-gas law, we found $\epsilon_{avg} = \frac{3}{2}k_BT$. By equating these expressions we find that the rms speed of molecules in a gas is

$$v_{rms} = \sqrt{\frac{3k_BT}{m}} \qquad (18.26)$$

The rms speed depends on the square root of the temperature and inversely on the square root of the molecular mass.

EXAMPLE 18.5 Calculating an rms speed

What is the rms speed of nitrogen molecules at room temperature (20°C)?

SOLVE The molecular mass is $m = 28$ u $= 4.68 \times 10^{-26}$ kg and $T = 20°C = 293$ K. It is then a simple calculation to find

$$v_{rms} = \sqrt{\frac{3(1.38 \times 10^{-23} \text{ J/K})(293 \text{ K})}{4.68 \times 10^{-26} \text{ kg}}} = 509 \text{ m/s}$$

Some speeds will be greater than this and others smaller, but 509 m/s will be a typical or fairly average speed. This is in excellent agreement with the experimental results of Figure 18.2.

EXAMPLE 18.6 Laser cooling

It is possible to "cool" atoms by letting them interact with a laser beam under proper, carefully controlled conditions. Laser cooling is currently a subject of intense research activity, and it is now possible to cool a dilute gas of atoms to a temperature lower than one *micro*kelvin! (The atoms are kept from solidifying by their extremely low density.) Various novel quantum effects appear at these incredibly low temperatures. What is the rms speed for cesium atoms at a temperature of 1.0 μK?

SOLVE Reference to the periodic table of the elements shows that the mass of a cesium atom is $m = 133$ u $= 2.22 \times 10^{-25}$ kg. At $T = 1.0 \mu K = 1.0 \times 10^{-6}$ K the rms speed is

$$v_{rms} = \sqrt{\frac{3(1.38 \times 10^{-23} \text{ J/K})(1.0 \times 10^{-6} \text{ K})}{2.22 \times 10^{-25} \text{ kg}}}$$

$$= 0.014 \text{ m/s} = 1.4 \text{ cm/s}$$

This is slow enough to enable us to "watch" the atoms moving about!

EXAMPLE 18.7 Mean time between collisions

Estimate the mean time between collisions for a nitrogen molecule at 1.0 atm pressure and room temperature (20°C).

MODEL Because v_{rms} is essentially the average molecular speed, the *mean time between collisions* is simply the time needed to travel distance λ, the mean free path, at speed v_{rms}.

SOLVE We found $\lambda = 2.3 \times 10^{-7}$ m in Example 18.1 and $v_{rms} = 509$ m/s in Example 18.5. Thus the mean time between collisions is

$$\tau_{coll} = \frac{\lambda}{v_{rms}} = \frac{2.3 \times 10^{-7} \text{ m}}{509 \text{ m/s}} = 4.5 \times 10^{-10} \text{ s}$$

ASSESS The air molecules around us move very fast, they collide with their neighbors about two billion times every second, and they manage to move, on average, only about 225 nm between collisions.

STOP TO THINK 18.3 Which system (or systems) has the largest average translational kinetic energy per molecule?

a. 1 mol of He at $p = 1$ atm, $T = 300$ K
b. 2 mol of He at $p = 2$ atm, $T = 300$ K
c. 1 mol of N_2 at $p = 0.5$ atm, $T = 600$ K
d. 2 mol of N_2 at $p = 0.5$ atm, $T = 450$ K
e. 1 mol of Ar at $p = 0.5$ atm, $T = 450$ K
f. 2 mol of Ar at $p = 2$ atm, $T = 300$ K

18.4 Thermal Energy and Specific Heat

We defined the thermal energy of a system to be $E_{th} = K_{micro} + U_{micro}$, where K_{micro} is the microscopic kinetic energy of the moving molecules and U_{micro} is the potential energy of the stretched and compressed molecular bonds. We're now ready to take a microscopic look at thermal energy. In doing so, we'll be able to resolve the puzzle of the molar specific heats.

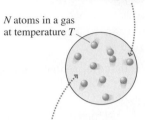

FIGURE 18.10 The atoms in a monatomic gas have only translational kinetic energy.

Atom i has translational kinetic energy ϵ_i but no potential energy or rotational kinetic energy.

N atoms in a gas at temperature T

The thermal energy of the gas is $E_{th} = \epsilon_1 + \epsilon_2 + \epsilon_3 + \cdots = N\epsilon_{avg}$.

Monatomic Gases

FIGURE 18.10 shows a monatomic gas such as helium or neon. The atoms in an ideal gas have no molecular bonds with their neighbors; hence $U_{micro} = 0$. Furthermore, the kinetic energy of a monatomic gas particle is entirely translational kinetic energy ϵ. Thus the thermal energy of a monatomic gas of N atoms is

$$E_{th} = K_{micro} = \epsilon_1 + \epsilon_2 + \epsilon_3 + \cdots + \epsilon_N = N\epsilon_{avg} \qquad (18.27)$$

where ϵ_i is the translational kinetic energy of atom i. We found that $\epsilon_{avg} = \frac{3}{2}k_B T$; hence the thermal energy is

$$E_{th} = \frac{3}{2}Nk_B T = \frac{3}{2}nRT \qquad \text{(thermal energy of a monatomic gas)} \qquad (18.28)$$

where we used $N = nN_A$ and the definition of Boltzmann's constant, $k_B = R/N_A$.

We've noted for the last two chapters that thermal energy is associated with temperature. Now we have an explicit result for a monatomic gas: E_{th} is directly proportional to the temperature. Notice that E_{th} is independent of the atomic mass. Any two monatomic gases will have the same thermal energy if they have the same temperature and the same number of atoms (or moles).

If the temperature of a monatomic gas changes by ΔT, its thermal energy changes by

$$\Delta E_{th} = \frac{3}{2}nR\Delta T \qquad (18.29)$$

In Chapter 17 we found that the change in thermal energy for *any* ideal gas process is related to the molar specific heat at constant volume by

$$\Delta E_{th} = nC_V\Delta T \qquad (18.30)$$

Equation 18.29 is a microscopic result that we obtained by relating the temperature to the average translational kinetic energy of the atoms. Equation 18.30 is a macroscopic result that we arrived at from the first law of thermodynamics. We can make a micro/macro connection by combining these two equations. Doing so gives us a *prediction* for the molar specific heat:

$$C_V = \frac{3}{2}R = 12.5 \text{ J/mol K} \qquad \text{(monatomic gas)} \qquad (18.31)$$

This was exactly the value of C_V for all three monatomic gases in Table 17.4. Not only has the micro/macro connection shown that C_V is the same for all monatomic gases—the puzzle we had noted in Chapter 17—it has also predicted the *value* of C_V. The perfect agreement of theory and experiment is strong evidence that gases really do consist of moving, colliding molecules.

The Equipartition Theorem

The particles of a monatomic gas are atoms. Their energy consists exclusively of their translational kinetic energy. A particle's translational kinetic energy can be written

$$\epsilon = \frac{1}{2}mv^2 = \frac{1}{2}mv_x^2 + \frac{1}{2}mv_y^2 + \frac{1}{2}mv_z^2 = \epsilon_x + \epsilon_y + \epsilon_z \qquad (18.32)$$

where we have written separately the energy associated with translational motion along the three axes. Because each axis in space is independent, we can think of ϵ_x, ϵ_y, and ϵ_z as independent *modes* of storing energy within the system.

Other systems have additional modes of energy storage. For example,

- Two atoms joined by a spring-like molecular bond can vibrate back and forth. Both kinetic and potential energy are associated with this vibration.
- A diatomic molecule, in addition to translational kinetic energy, has rotational kinetic energy if it rotates end-over-end like a dumbbell.

It will be useful to define the number of **degrees of freedom** as the number of distinct and independent modes of energy storage. A monatomic gas has three degrees of freedom, the three modes of translational kinetic energy. Systems that can vibrate or rotate have more degrees of freedom.

An important result of statistical physics says that the energy in a system is distributed so that all modes of energy storage have equal amounts of energy. This conclusion is known as the *equipartition theorem,* meaning that the energy is equally divided. The proof is beyond what we can do in this textbook, so we will state the theorem without proof:

> **Equipartition theorem** The thermal energy of a system of particles is equally divided among all the possible energy modes. For a system of N particles at temperature T, the energy stored in each mode (each degree of freedom) is $\frac{1}{2}Nk_BT$ or, in terms of moles, $\frac{1}{2}nRT$.

A monatomic gas has three degrees of freedom and thus, as we found above, $E_{th} = \frac{3}{2}Nk_BT$.

Solids

FIGURE 18.11 reminds you of our "bedspring model" of a solid with particle-like atoms connected by a lattice of spring-like molecular bonds. How many degrees of freedom does a solid have? The kinetic energy of an atom as it vibrates around its equilibrium position is given by Equation 18.32. Three degrees of freedom are associated with the kinetic energy, just as in a monatomic gas. In addition, the molecular bonds can be compressed or stretched independently along the x-, y-, and z-axes. Three additional degrees of freedom are associated with these three modes of potential energy. Altogether, a solid has six degrees of freedom.

The energy stored in each of these six degrees of freedom is $\frac{1}{2}Nk_BT$. The thermal energy of a solid is the total energy stored in all six modes, or

$$E_{th} = 3Nk_BT = 3nRT \qquad \text{(thermal energy of a solid)} \qquad (18.33)$$

We can use this result to predict the molar specific heat of a solid. If the temperature changes by ΔT, then the thermal energy changes by

$$\Delta E_{th} = 3nR\Delta T \qquad (18.34)$$

In Chapter 17 we defined the molar specific heat of a solid such that

$$\Delta E_{th} = nC\Delta T \qquad (18.35)$$

By comparing Equations 18.34 and 18.35 we can predict that the molar specific heat of a solid is

$$C = 3R = 25.0 \text{ J/mol K} \qquad \text{(solid)} \qquad (18.36)$$

Not bad. The five elemental solids in Table 17.2 had molar specific heats clustered right around 25 J/mol K. They ranged from 24.3 J/mol K for aluminum to 26.5 J/mol K for lead. There are two reasons the agreement between theory and experiment isn't quite as perfect as it was for monatomic gases. First, our simple bedspring model of a solid isn't quite as accurate as our model of a monatomic gas. Second, quantum effects are beginning to make their appearance. More on this shortly. Nonetheless, our

FIGURE 18.11 A simple model of a solid.

Each atom has microscopic translational kinetic energy *and* microscopic potential energy along all three axes.

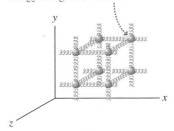

ability to predict C to within a few percent from a simple model of a solid is further evidence for the atomic structure of matter.

Diatomic Molecules

Diatomic molecules are a bigger challenge. How many degrees of freedom does a diatomic molecule have? FIGURE 18.12 shows a diatomic molecule, such as molecular nitrogen N_2, oriented along the x-axis. Three degrees of freedom are associated with the molecule's translational kinetic energy. The molecule can have a dumbbell-like end-over-end rotation about either the y-axis or the z-axis. It can also rotate about its own axis. These are three rotational degrees of freedom. The two atoms can also vibrate back and forth, stretching and compressing the molecular bond. This vibrational motion has both kinetic and potential energy—thus two more degrees of freedom.

Altogether, then, a diatomic molecule has eight degrees of freedom, and we would expect the thermal energy of a gas of diatomic molecules to be $E_{th} = 4k_B T$. The analysis we followed for a monatomic gas would then lead to the prediction $C_V = 4R = 33.2$ J/mol K. As compelling as this reasoning seems to be, this is *not* the experimental value of C_V that was reported for diatomic gases in Table 17.4. Instead, we found $C_V = 20.8$ J/mol K.

Why should a theory that works so well for monatomic gases and solids fail so miserably for diatomic molecules? To see what's going on, notice that 20.8 J/mol K = $\frac{5}{2}R$. A monatomic gas, with three degrees of freedom, has $C_V = \frac{3}{2}R$. A solid, with six degrees of freedom, has $C = 3R$. A diatomic gas would have $C_V = \frac{5}{2}R$ if it had five degrees of freedom, not eight.

This discrepancy was a major conundrum as statistical physics developed in the late 19th century. Although it was not recognized as such at the time, we are here seeing our first evidence for the breakdown of classical Newtonian physics. Classically, a diatomic molecule has eight degrees of freedom. The equipartition theorem doesn't distinguish between them; all eight should have the same energy. But atoms and molecules are not classical particles. It took the development of quantum theory in the 1920s to accurately characterize the behavior of atoms and molecules. We don't yet have the tools needed to see why, but quantum effects prevent three of the modes—the two vibrational modes and the rotation of the molecule about its own axis—from being active at room temperature.

FIGURE 18.13 shows C_V as a function of temperature for hydrogen gas. C_V is right at $\frac{5}{2}R$ for temperatures from ≈ 200 K up to ≈ 800 K. But at very low temperatures C_V drops to the monatomic-gas value $\frac{3}{2}R$. The two rotational modes become "frozen out" and the nonrotating molecule has only translational kinetic energy. Quantum physics can explain this, but not Newtonian physics. You can also see that the two vibrational modes *do* become active at very high temperatures, where C_V rises to $\frac{7}{2}R$. Thus the real answer to "What's wrong?" is that Newtonian physics is not the right physics for

FIGURE 18.12 A diatomic molecule can rotate or vibrate.

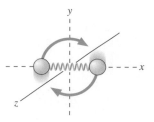

Rotation end-over-end about the z-axis

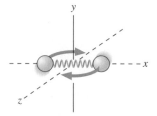

Rotation end-over-end about the y-axis

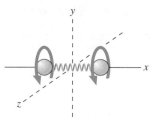

Rotation about the x-axis

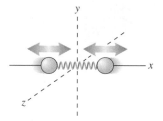

Vibration back and forth along the x-axis

FIGURE 18.13 Hydrogen molar specific heat at constant volume as a function of temperature. The temperature scale is logarithmic.

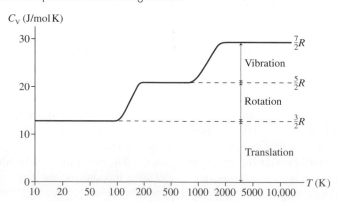

describing atoms and molecules. We are somewhat fortunate that Newtonian physics is adequate to understand monatomic gases and solids, at least at room temperature.

Accepting the quantum result that a diatomic gas has only five degrees of freedom at commonly used temperatures (the translational degrees of freedom and the two end-over-end rotations), we find

$$E_{th} = \frac{5}{2}Nk_BT = \frac{5}{2}nRT$$

(diatomic gases) (18.37)

$$C_V = \frac{5}{2}R = 20.8 \text{ J/mol K}$$

A diatomic gas has more thermal energy than a monatomic gas at the same temperature because the molecules have rotational as well as translational kinetic energy.

While the micro/macro connection firmly establishes the atomic structure of matter, it also heralds the need for a new theory of matter at the atomic level. That is a task we will take up in Part VII. For now, Table 18.2 summarizes what we have learned from kinetic theory about thermal energy and molar specific heats.

TABLE 18.2 Kinetic theory predictions for the thermal energy and the molar specific heat

System	Degrees of freedom	E_{th}	C_V
Monatomic gas	3	$\frac{3}{2}Nk_BT = \frac{3}{2}nRT$	$\frac{3}{2}R = 12.5$ J/mol K
Diatomic gas	5	$\frac{5}{2}Nk_BT = \frac{5}{2}nRT$	$\frac{5}{2}R = 20.8$ J/mol K
Elemental solid	6	$3Nk_BT = 3nRT$	$3R = 25.0$ J/mol K

EXAMPLE 18.8 The rotational frequency of a molecule
The nitrogen molecule N_2 has a bond length of 0.12 nm. Estimate the rotational frequency of N_2 at 20°C.

MODEL The molecule can be modeled as a rigid dumbbell of length $L = 0.12$ nm rotating about its center.

SOLVE The rotational kinetic energy of the molecule is $\epsilon_{rot} = \frac{1}{2}I\omega^2$, where I is the moment of inertia about the center. Because we have two point masses each moving in a circle of radius $r = L/2$, the moment of inertia is

$$I = mr^2 + mr^2 = 2m\left(\frac{L}{2}\right)^2 = \frac{mL^2}{2}$$

Thus the rotational kinetic energy is

$$\epsilon_{rot} = \frac{1}{2}\frac{mL^2}{2}\omega^2 = \frac{mL^2\omega^2}{4} = \pi^2 mL^2 f^2$$

where we used $\omega = 2\pi f$ to relate the rotational frequency f to the angular frequency ω. From the equipartition theorem, the energy

associated with this mode is $\frac{1}{2}Nk_BT$, so the *average* rotational kinetic energy per molecule is

$$(\epsilon_{rot})_{avg} = \frac{1}{2}k_BT$$

Equating these two expressions for ϵ_{rot} gives us

$$\pi^2 mL^2 f^2 = \frac{1}{2}k_BT$$

Thus the rotational frequency is

$$f = \sqrt{\frac{k_BT}{2\pi^2 mL^2}} = 7.8 \times 10^{11} \text{ rev/s}$$

We evaluated f at $T = 293$ K, using $m = 14$ u $= 2.34 \times 10^{-26}$ kg for each *atom*.

ASSESS This is a *very* high frequency, but these values are typical of molecular rotations.

STOP TO THINK 18.4 How many degrees of freedom does a bead on a rigid rod have?

a. 1   b. 2   c. 3   d. 4   e. 5   f. 6

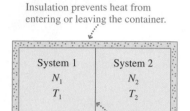

Insulation prevents heat from entering or leaving the container.

System 1
N_1
T_1

System 2
N_2
T_2

A thin barrier prevents atoms from moving from system 1 to 2 but still allows them to collide. The barrier is clamped in place and cannot move.

18.5 Thermal Interactions and Heat

We can now look in more detail at what happens when two systems at different temperatures interact with each other. **FIGURE 18.14** shows a rigid, insulated container divided into two sections by a very thin, stiff membrane. The left side, which we'll call system 1, has N_1 atoms at an initial temperature T_{1i}. System 2 on the right has N_2 atoms at an initial temperature T_{2i}. The membrane is so thin that atoms can collide at the boundary as if the membrane were not there, yet it is a barrier that prevents atoms from moving from one side to the other. The situation is analogous, on an atomic scale, to basketballs colliding through a shower curtain.

Suppose that system 1 is initially at a higher temperature: $T_{1i} > T_{2i}$. This is not an equilibrium situation. The temperatures will change with time until the systems eventually reach a common final temperature T_f. If you *watch* the gases as one warms and the other cools, you see nothing happening. This interaction is quite different from a mechanical interaction in which, for example, you might see a piston move from one side toward the other. The only way in which the gases can interact is via molecular collisions at the boundary. This is a *thermal interaction,* and our goal is to understand how thermal interactions bring the systems to thermal equilibrium.

System 1 and system 2 begin with thermal energies

$$E_{1i} = \frac{3}{2}N_1 k_B T_{1i} = \frac{3}{2}n_1 R T_{1i}$$

$$E_{2i} = \frac{3}{2}N_2 k_B T_{2i} = \frac{3}{2}n_2 R T_{2i}$$

(18.38)

We've written the energies for monatomic gases; you could do the same calculation if one or both of the gases is diatomic by replacing the $\frac{3}{2}$ with $\frac{5}{2}$. Notice that we've omitted the subscript "th" to keep the notation manageable.

The total energy of the combined systems is $E_{tot} = E_{1i} + E_{2i}$. As systems 1 and 2 interact, their individual thermal energies E_1 and E_2 can change but their sum E_{tot} remains constant. The system will have reached thermal equilibrium when the individual thermal energies reach final values E_{1f} and E_{2f} that no longer change.

The Systems Exchange Energy

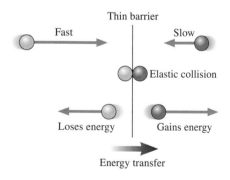

Thin barrier

Fast

Slow

Elastic collision

Loses energy

Gains energy

Energy transfer

FIGURE 18.15 shows a fast atom and a slow atom approaching the barrier from opposite sides. They undergo a perfectly elastic collision at the barrier. Although no net energy is lost in a perfectly elastic collision, the faster atom loses energy while the slower one gains energy. In other words, there is an energy *transfer* from the faster atom's side to the slower atom's side.

The average translational kinetic energy per molecule is directly proportional to the temperature: $\epsilon_{avg} = \frac{3}{2}k_B T$. Because $T_{1i} > T_{2i}$, the atoms in system 1 are, on average, more energetic than the atoms in system 2. Thus *on average* the collisions transfer energy from system 1 to system 2. Not in every collision: sometimes a fast atom in system 2 collides with a slow atom in system 1, transferring energy from 2 to 1. But the net energy transfer, from all collisions, is from the warmer system 1 to the cooler system 2. In other words, **heat is the energy transferred *via collisions* between the more energetic (warmer) atoms on one side and the less energetic (cooler) atoms on the other.**

How do the systems "know" when they've reached thermal equilibrium? Energy transfer continues until the atoms on both sides of the barrier have the *same average translational kinetic energy*. Once the average translational kinetic energies are the same, there is no tendency for energy to flow in either direction. This is the state of thermal equilibrium, so the condition for thermal equilibrium is

$$(\epsilon_1)_{avg} = (\epsilon_2)_{avg} \quad \text{(thermal equilibrium)}$$

(18.39)

where, as before, ϵ is the translational kinetic energy of an atom.

Because the average energies are directly proportional to the final temperatures, $\epsilon_{avg} = \frac{3}{2}k_B T_f$, thermal equilibrium is characterized by the macroscopic condition

$$T_{1f} = T_{2f} = T_f \qquad \text{(thermal equilibrium)} \qquad (18.40)$$

In other words, **two thermally interacting systems reach a common final temperature** *because* **they exchange energy via collisions until the atoms on each side have, on average, equal translational kinetic energies.** This is a very important idea.

Equation 18.40 can be used to determine the equilibrium thermal energies. Because these are monatomic gases, $E_{th} = N\epsilon_{avg}$. Thus the equilibrium condition $(\epsilon_1)_{avg} = (\epsilon_2)_{avg} = (\epsilon_{tot})_{avg}$ implies

$$\frac{E_{1f}}{N_1} = \frac{E_{2f}}{N_2} = \frac{E_{tot}}{N_1 + N_2} \qquad (18.41)$$

from which we can conclude

$$E_{1f} = \frac{N_1}{N_1 + N_2}E_{tot}$$

$$E_{2f} = \frac{N_2}{N_1 + N_2}E_{tot} \qquad (18.42)$$

This result can also be written in terms of the number of moles. If we use $N = N_A n$ and note that the N_A cancels, Equation 18.42 becomes

$$E_{1f} = \frac{n_1}{n_1 + n_2}E_{tot}$$

$$E_{2f} = \frac{n_2}{n_1 + n_2}E_{tot} \qquad (18.43)$$

Notice that $E_{1f} + E_{2f} = E_{tot}$, verifying that energy has been conserved even while being redistributed between the systems.

No work is done on either system because the barrier has no macroscopic displacement, so the first law of thermodynamics is

$$Q_1 = \Delta E_1 = E_{1f} - E_{1i}$$

$$Q_2 = \Delta E_2 = E_{2f} - E_{2i} \qquad (18.44)$$

As a homework problem you can show that $Q_1 = -Q_2$, as required by energy conservation. That is, the heat lost by one system is gained by the other. $|Q_1|$ is the quantity of heat that is transferred from the warmer gas to the cooler gas during the thermal interaction.

NOTE ▶ In general, the equilibrium thermal energies of the system are *not* equal. That is, $E_{1f} \neq E_{2f}$. They will be equal only if $N_1 = N_2$. Equilibrium is reached when the average translational kinetic energies in the two systems are equal—that is, when $(\epsilon_1)_{avg} = (\epsilon_2)_{avg}$, not when $E_{1f} = E_{2f}$. The distinction is important. FIGURE 18.16 summarizes these ideas. ◀

FIGURE 18.16 Equilibrium is reached when the atoms on each side have, on average, equal energies.

Collisions transfer energy from the warmer system to the cooler system as more energetic atoms lose energy to less energetic atoms.

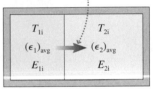

Thermal equilibrium occurs when the systems have the same average translational kinetic energy and thus the same temperature.

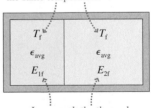

In general, the thermal energies E_{1f} and E_{2f} are *not* equal.

EXAMPLE 18.9 A thermal interaction

A sealed, insulated container has 2.0 g of helium at an initial temperature of 300 K on one side of a barrier and 10.0 g of argon at an initial temperature of 600 K on the other side.

a. How much heat energy is transferred, and in which direction?
b. What is the final temperature?

MODEL The systems start with different temperatures, so they are not in thermal equilibrium. Energy will be transferred via collisions from the argon to the helium until both systems have the same average molecular energy.

SOLVE a. Let the helium be system 1. Helium has molar mass $M_{mol} = 4$ g/mol, so $n_1 = M/M_{mol} = 0.50$ mol.

Continued

Similarly, argon has $M_{mol} = 40$ g/mol, so $n_2 = 0.25$ mol. The initial thermal energies of the two monatomic gases are

$$E_{1i} = \frac{3}{2}n_1 RT_{1i} = 225R = 1870 \text{ J}$$

$$E_{2i} = \frac{3}{2}n_2 RT_{2i} = 225R = 1870 \text{ J}$$

The systems start with *equal* thermal energies, but they are not in thermal equilibrium. The total energy is $E_{tot} = 3740$ J. In equilibrium, this energy is distributed between the two systems as

$$E_{1f} = \frac{n_1}{n_1 + n_2}E_{tot} = \frac{0.50}{0.75}3740 \text{ J} = 2493 \text{ J}$$

$$E_{2f} = \frac{n_2}{n_1 + n_2}E_{tot} = \frac{0.25}{0.75}3740 \text{ J} = 1247 \text{ J}$$

The heat entering or leaving each system is

$$Q_1 = Q_{He} = E_{1f} - E_{1i} = 623 \text{ J}$$

$$Q_2 = Q_{Ar} = E_{2f} - E_{2i} = -623 \text{ J}$$

The helium and the argon interact thermally via collisions at the boundary, causing 623 J of heat to be transferred from the warmer argon to the cooler helium.

b. These are constant-volume processes, thus $Q = nC_V\Delta T$. $C_V = \frac{3}{2}R$ for monatomic gases, so the temperature changes are

$$\Delta T_{He} = \frac{Q_{He}}{\frac{3}{2}nR} = \frac{623 \text{ J}}{1.5(0.50 \text{ mol})(8.31 \text{ J/mol K})} = 100 \text{ K}$$

$$\Delta T_{Ar} = \frac{Q_{Ar}}{\frac{3}{2}nR} = \frac{-623 \text{ J}}{1.5(0.25 \text{ mol})(8.31 \text{ J/mol K})} = -200 \text{ K}$$

Both gases reach the common final temperature $T_f = 400$ K.

ASSESS $E_{1f} = 2E_{2f}$ because there are twice as many atoms in system 1.

The main idea of this section is that two systems reach a common final temperature not by magic or by a prearranged agreement but simply from the energy exchange of vast numbers of molecular collisions. Real interacting systems, of course, are separated by walls rather than our unrealistic thin membrane. As the systems interact, the energy is first transferred via collisions from system 1 into the wall and subsequently, as the cooler molecules collide with a warm wall, into system 2. That is, the energy transfer is $E_1 \rightarrow E_{wall} \rightarrow E_2$. This is still heat because the energy transfer is occurring via molecular collisions rather than mechanical motion.

STOP TO THINK 18.5 Systems A and B are interacting thermally. At this instant of time,

a. $T_A > T_B$
b. $T_A = T_B$
c. $T_A < T_B$

A	B
$N = 1000$	$N = 2000$
$\epsilon_{avg} = 1.0 \times 10^{-20}$ J	$\epsilon_{avg} = 0.5 \times 10^{-20}$ J
$E_{th} = 1.0 \times 10^{-17}$ J	$E_{th} = 1.0 \times 10^{-17}$ J

18.6 Irreversible Processes and the Second Law of Thermodynamics

The preceding section looked at the thermal interaction between a warm gas and a cold gas. Heat energy is transferred from the warm gas to the cold gas until they reach a common final temperature. But why isn't heat transferred from the cold gas to the warm gas, making the cold side colder and the warm side warmer? Such a process could still conserve energy, but it never happens. The transfer of heat energy from hot to cold is an example of an **irreversible process**, a process that can happen only in one direction.

Examples of irreversible processes abound. Stirring the cream in your coffee mixes the cream and coffee together. No amount of stirring ever unmixes them. If you shake a jar that has red marbles on the top and blue marbles on the bottom, the two colors are quickly mixed together. No amount of shaking ever separates them again. If you watched a movie of someone shaking a jar and saw the red and blue marbles separating, you would be certain that the movie was running backward. In fact, a reasonable definition of an irreversible process is one for which a backward-running movie shows a physically impossible process.

FIGURE 18.17a is a two-frame movie of a collision between two particles, perhaps two gas molecules. Suppose that sometime after the collision is over we could reach in and reverse the velocities of both particles. That is, replace vector $\vec{v}$ with vector $-\vec{v}$. Then, as in a movie playing backward, the collision would happen in reverse. This is the movie of **FIGURE 18.17b**.

FIGURE 18.17 Molecular collisions are reversible.

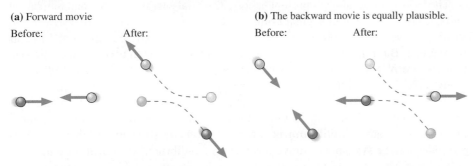

(a) Forward movie

Before: After:

(b) The backward movie is equally plausible.

Before: After:

You cannot tell, just by looking at the two movies, which is really going forward and which is being played backward. Maybe Figure 18.17b was the original collision and Figure 18.17a is the backward version. Nothing in either collision looks wrong, and no measurements you might make on either would reveal any violations of Newton's laws. Interactions at the molecular level are reversible processes.

Contrast this with the two-frame car crash movies in **FIGURE 18.18**. Past and future are clearly distinct in an irreversible process, and the backward movie of **FIGURE 18.18b** is obviously wrong. But what has been violated in the backward movie? To have the crumpled car spring away from the wall would not violate any laws of physics we have so far discovered. It would simply require transforming the thermal energy of the car and wall back into the macroscopic center-of-mass energy of the car as a whole.

The paradox stems from our assertion that macroscopic phenomena can be understood on the basis of microscopic molecular motions. If the microscopic motions are all reversible, how can the macroscopic phenomena end up being irreversible? If reversible collisions can cause heat to be transferred from hot to cold, why do they never cause heat to be transferred from cold to hot? There must be another law of physics preventing it. The law we seek must, in some sense, be able to distinguish the past from the future.

FIGURE 18.18 A car crash is irreversible.

(a) Forward movie

Before: After:

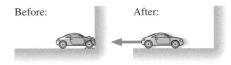

(b) The backward movie is physically impossible.

Before: After:

Which Way to Equilibrium?

Stated another way, how do two systems initially at different temperatures "know" which way to go to reach equilibrium? Perhaps an analogy will help.

FIGURE 18.19 shows two boxes, numbered 1 and 2, containing identical balls. Box 1 starts with more balls than box 2, so $N_{1i} > N_{2i}$. Once every second, one ball is chosen at random and moved to the other box. This is a reversible process because a ball can move from box 2 to box 1 just as easily as from box 1 to box 2. What do you expect to see if you return several hours later?

Because balls are chosen at random, and because $N_{1i} > N_{2i}$, it's initially more likely that a ball will move from box 1 to box 2 than from box 2 to box 1. Sometimes a ball will move "backward" from box 2 to box 1, but overall there's a net movement of balls from box 1 to box 2. The system will evolve until $N_1 \approx N_2$. This is a stable situation—equilibrium!—with an equal number of balls moving in both directions.

But couldn't it go the other way, with N_1 getting even larger while N_2 decreases? In principle, any possible arrangement of the balls is possible in the same way that any

FIGURE 18.19 Two interacting systems. Balls are chosen at random and moved to the other box.

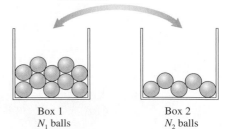

Balls are chosen at random and moved from one box to the other.

Box 1
N_1 balls

Box 2
N_2 balls

number of heads are possible if you throw N coins in the air and let them fall. If you throw four coins, the odds are 1 in 2^4, or 1 in 16, of getting four heads. With four balls, the odds are 1 in 16 that, at a randomly chosen instant of time, you would find $N_1 = 4$. You wouldn't find that to be terribly surprising.

With 10 balls, the probability that $N_1 = 10$ is $0.5^{10} \approx 1/1000$. With 100 balls, the probability that $N_1 = 100$ has dropped to $\approx 10^{-30}$. With 10^{20} balls, the odds of finding all of them, or even most of them, in one box are so staggeringly small that it's safe to say it will "never" happen. Although each transfer is reversible, **the statistics of large numbers make it overwhelmingly more likely that the system will evolve toward a state in which $N_1 \approx N_2$ than toward a state in which $N_1 > N_2$.**

The balls in our analogy represent energy. The total energy, like the total number of balls, is conserved, but molecular collisions can move energy between system 1 and system 2. Each collision is reversible, just as likely to transfer energy from 1 to 2 as from 2 to 1. But if $(\epsilon_{1i})_{avg} > (\epsilon_{2i})_{avg}$, and if we're dealing with two macroscopic systems where $N > 10^{20}$, then it's overwhelmingly likely that the net result of many, many collisions will be to transfer energy from system 1 to system 2 until $(\epsilon_{1f})_{avg} = (\epsilon_{2f})_{avg}$—in other words, for heat energy to be transferred from hot to cold.

The system reaches thermal equilibrium not by any plan or by outside intervention, but simply because **equilibrium is the *most probable* state in which to be.** It is *possible* that the system will move away from equilibrium, with heat moving from cold to hot, but remotely improbable in any realistic system. The consequence of a vast number of random events is that the system evolves in one direction, toward equilibrium, and not the other. **Reversible microscopic events lead to irreversible macroscopic behavior because some macroscopic states are vastly more probable than others.**

Order, Disorder, and Entropy

FIGURE 18.20 shows three different systems. At the top is a group of atoms arranged in a crystal-like lattice. This is a highly ordered and nonrandom system, with each atom's position precisely specified. Contrast this with the system on the bottom, where there is no order at all. The position of every atom was assigned entirely at random.

It is extremely improbable that the atoms in a container would *spontaneously* arrange themselves into the ordered pattern of the top picture. In a system of, say, 10^{20} atoms, the probability of this happening is similar to the probability that 10^{20} tossed coins will all be heads. We can safely say that it will never happen. By contrast, there are a vast number of arrangements like the one on the bottom that randomly fill the container.

The middle picture of Figure 18.20 is an in-between situation. This situation might arise as a solid melts. The positions of the atoms are clearly not completely random, so the system preserves some degree of order. This in-between situation is more likely to occur spontaneously than the highly ordered lattice on the top, but is less likely to occur than the completely random system on the bottom.

Scientists and engineers use a state variable called **entropy** to measure the probability that a macroscopic state will occur spontaneously. The ordered lattice, which has a very small probability of spontaneous occurrence, has a very low entropy. The entropy of the randomly filled container is high. The entropy of the middle picture is somewhere in between. It is often said that entropy measures the amount of *disorder* in a system. The entropy in Figure 18.20 increases as you move from the ordered system on the top to the disordered system on the bottom.

Similarly, two thermally interacting systems with different temperatures have a low entropy. These systems are ordered in the sense that the faster atoms are on one side of the barrier, the slower atoms on the other. The most random possible distribution of energy, and hence the least ordered system, corresponds to the situation where the two systems are in thermal equilibrium with equal temperatures. Entropy increases as two systems with initially different temperatures move toward equilibrium.

FIGURE 18.20 Ordered and disordered systems.

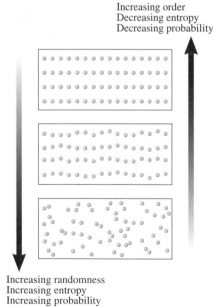

Increasing order
Decreasing entropy
Decreasing probability

Increasing randomness
Increasing entropy
Increasing probability

Entropy would decrease if heat energy moved from cold to hot, making the hot system hotter and the cold system colder.

Entropy can be calculated, but we'll leave that to more advanced courses. For our purposes, the *concept* of entropy as a measure of the disorder in a system, or of the probability that a macroscopic state will occur, is more important than a numerical value.

The Second Law of Thermodynamics

The fact that macroscopic systems evolve irreversibly toward equilibrium is a statement about nature that is not contained in any of the laws of physics we have encountered. It is, in fact, a new law of physics, one known as the **second law of thermodynamics.**

The formal statement of the second law of thermodynamics is given in terms of entropy:

Tossing all heads, while not impossible, is extremely unlikely, and the probability of doing so rapidly decreases as the number of coins increases.

> **Second law, formal statement** The entropy of an isolated system (or group of systems) never decreases. The entropy either increases, until the system reaches equilibrium, or, if the system began in equilibrium, stays the same.

The qualifier "isolated" is most important. We can order the system by reaching in from the outside, perhaps using tiny tweezers to place the atoms in a lattice. Similarly, we can transfer heat from cold to hot by using a refrigerator. The second law is about what a system can or cannot do *spontaneously,* on its own, without outside intervention.

The second law of thermodynamics tells us that an isolated system evolves such that

- Order turns into disorder and randomness.
- Information is lost rather than gained.
- The system "runs down."

An isolated system never spontaneously generates order out of randomness. It is not that the system "knows" about order or randomness, but rather that there are vastly more states corresponding to randomness than there are corresponding to order. As collisions occur at the microscopic level, the laws of probability dictate that the system will, on average, move inexorably toward the most probable and thus most random macroscopic state.

The second law of thermodynamics is often stated in several equivalent but more informal versions. One of these, and the one most relevant to our discussion, is

> **Second law, informal statement #1** When two systems at different temperatures interact, heat energy is transferred spontaneously from the hotter to the colder system, never from the colder to the hotter.

The second law of thermodynamics is an independent statement about nature, separate from the first law. The first law is a precise statement about energy conservation. The second law, by contrast, is a *probabilistic* statement, based on the statistics of very large numbers. While it is conceivable that heat could spontaneously move from cold to hot, it will never occur in any realistic macroscopic system.

The irreversible evolution from less-likely macroscopic states to more-likely macroscopic states is what gives us a macroscopic direction of time. Stirring blends your coffee and cream, it never unmixes them. Friction causes an object to stop while increasing its thermal energy; the random atomic motions of thermal energy never spontaneously organize themselves into a macroscopic motion of the entire object. A plant in a sealed jar dies and decomposes to carbon and various gases; the gases and carbon never spontaneously assemble themselves into a flower. These are all examples of irreversible processes. They each show a clear direction of time, a distinct difference between past and future.

Thus another statement of the second law is

> **Second law, informal statement #2**   The time direction in which the entropy of an isolated macroscopic system increases is "the future."

Establishing the "arrow of time" is one of the most profound implications of the second law of thermodynamics.

The second law of thermodynamics has important implications for issues ranging from how we as a society use energy and resources to biological evolution and the future of the universe. We'll return to some of these issues in the Summary to Part IV. In the meantime, the second law will be used in Chapter 19 to understand some of the practical aspects of the thermodynamics of engines.

STOP TO THINK 18.6   Two identical boxes each contain 1,000,000 molecules. In box A, 750,000 molecules happen to be in the left half of the box while 250,000 are in the right half. In box B, 499,900 molecules happen to be in the left half of the box while 500,100 are in the right half. At this instant of time,

a. The entropy of box A is larger than the entropy of box B.
b. The entropy of box A is equal to the entropy of box B.
c. The entropy of box A is smaller than the entropy of box B.

SUMMARY

The goal of Chapter 18 has been to understand the properties of a macroscopic system in terms of the microscopic behavior of its molecules.

General Principles

Kinetic theory, the **micro/macro connection,** relates the macroscopic properties of a system to the motion and collisions of its atoms and molecules.

The Equipartition Theorem

Tells us how collisions distribute the energy in the system. The energy stored in each mode of the system (each **degree of freedom**) is $\frac{1}{2}Nk_BT$ or, in terms of moles, $\frac{1}{2}nRT$.

The Second Law of Thermodynamics

Tells us how collisions move a system toward equilibrium. The entropy of an isolated system can only increase or, in equilibrium, stay the same.

• Order turns into disorder and randomness.

• Systems run down.

• Heat energy is transferred spontaneously from a hotter to a colder system, never from colder to hotter.

Important Concepts

Pressure is due to the force of the molecules colliding with the walls:

$$p = \frac{1}{3}\frac{N}{V}mv_{rms}^2 = \frac{2}{3}\frac{N}{V}\epsilon_{avg}$$

The **average translational kinetic energy** of a molecule is $\epsilon_{avg} = \frac{3}{2}k_BT$. The temperature of the gas $T = \frac{2}{3k_B}\epsilon_{avg}$ measures the average translational kinetic energy.

Entropy measures the probability that a macroscopic state will occur or, equivalently, the amount of disorder in a system.

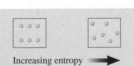

Increasing entropy ➡

The **thermal energy** of a system is

$$E_{th} = \text{translational kinetic energy} + \text{rotational kinetic energy} + \text{vibrational energy}$$

• **Monatomic gas** $E_{th} = \frac{3}{2}Nk_BT = \frac{3}{2}nRT$

• **Diatomic gas** $E_{th} = \frac{5}{2}Nk_BT = \frac{5}{2}nRT$

• **Elemental solid** $E_{th} = 3Nk_BT = 3nRT$

Heat is energy transferred via collisions from more-energetic molecules on one side to less-energetic molecules on the other. Equilibrium is reached when $(\epsilon_1)_{avg} = (\epsilon_2)_{avg}$, which implies $T_{1f} = T_{2f}$.

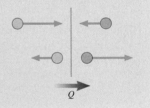

Applications

The **root-mean-square speed** v_{rms} is the square root of the average of the squares of the molecular speeds:

$$v_{rms} = \sqrt{(v^2)_{avg}}$$

For molecules of mass m at temperature T, $v_{rms} = \sqrt{\dfrac{3k_BT}{m}}$

Molar specific heats can be predicted from the thermal energy because $\Delta E_{th} = nC\Delta T$.

• **Monatomic gas** $C_V = \frac{3}{2}R$

• **Diatomic gas** $C_V = \frac{5}{2}R$

• **Elemental solid** $C = 3R$

Terms and Notation

kinetic theory
histogram
mean free path, λ

root-mean-square speed, v_{rms}
degrees of freedom
equipartition theorem

irreversible process
entropy
second law of thermodynamics

(MP)   For homework assigned on MasteringPhysics, go to
www.masteringphysics.com

Problem difficulty is labeled as | (straightforward) to ||| (challenging).

Problems labeled ■ integrate significant material from earlier chapters.

CONCEPTUAL QUESTIONS

1. Solids and liquids resist being compressed. They are not totally incompressible, but it takes large forces to compress them even slightly. If it is true that matter consists of atoms, what can you infer about the microscopic nature of solids and liquids from their incompressibility?

2. Gases, in contrast with solids and liquids, are very compressible. What can you infer from this observation about the microscopic nature of gases?

3. The density of air at STP is about $\frac{1}{1000}$ the density of water. How does the average distance between air molecules compare to the average distance between water molecules? Explain.

4. The mean free path of molecules in a gas is 200 nm.
   a. What will be the mean free path if the pressure is doubled while all other state variables are held constant?
   b. What will be the mean free path if the absolute temperature is doubled while all other state variables are held constant?

5. If the pressure of a gas is really due to the *random* collisions of molecules with the walls of the container, why do pressure gauges—even very sensitive ones—give perfectly steady readings? Shouldn't the gauge be continually jiggling and fluctuating? Explain.

6. Suppose you could suddenly increase the speed of every molecule in a gas by a factor of 2.
   a. Would the rms speed of the molecules increase by a factor of $2^{1/2}$, 2, or 2^2? Explain.
   b. Would the gas pressure increase by a factor of $2^{1/2}$, 2, or 2^2? Explain.

7. Suppose you could suddenly increase the speed of every molecule in a gas by a factor of 2.
   a. Would the temperature of the gas increase by a factor of $2^{1/2}$, 2, or 2^2? Explain.

   b. Would the molar specific heat at constant volume change? If so, by what factor? If not, why not?

8. The two containers of gas in FIGURE Q18.8 are in good thermal contact with each other but well insulated from the environment. They have been in contact for a long time and are in thermal equilibrium.

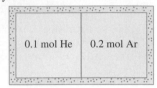

FIGURE Q18.8

   a. Is v_{rms} of helium greater than, less than, or equal to v_{rms} of argon? Explain.
   b. Does the helium have more thermal energy, less thermal energy, or the same amount of thermal energy as the argon? Explain.

9. Suppose you place an ice cube in a beaker of room-temperature water, then seal them in a rigid, well-insulated container. No energy can enter or leave the container.
   a. If you open the container an hour later, will you find a beaker of water slightly cooler than room temperature, or a large ice cube and some 100°C steam?
   b. Finding a large ice cube and some 100°C steam would not violate the first law of thermodynamics. $W = 0$ J and $Q = 0$ J because the container is sealed, and $\Delta E_{th} = 0$ J because the increase in thermal energy of the water molecules that became steam is offset by the decrease in thermal energy of the water molecules that turned to ice. Energy would be conserved, yet we never see an outcome like this. Why not?

EXERCISES AND PROBLEMS

Exercises

Section 18.1 Molecular Speeds and Collisions

1. | The number density of an ideal gas at STP is called the *Loschmidt number*. Calculate the Loschmidt number.

2. | A 1.0 m × 1.0 m × 1.0 m cube of nitrogen gas is at 20°C and 1.0 atm. Estimate the number of molecules in the cube with a speed between 700 m/s and 1000 m/s.

3. | At what pressure will the mean free path in room-temperature (20°C) nitrogen be 1.0 m?

4. || Integrated circuits are manufactured in vacuum chambers in which the air pressure is 1.0×10^{-10} mm of Hg. What are (a) the number density and (b) the mean free path of a molecule? Assume $T = 20°C$.

5. | The mean free path of a molecule in a gas is 300 nm. What will the mean free path be if the gas temperature is doubled at (a) constant volume and (b) constant pressure?

6. || The pressure inside a tank of neon is 150 atm. The temperature is 25°C. On average, how many atomic diameters does a neon atom move between collisions?

7. || A lottery machine uses blowing air to keep 2000 Ping-Pong balls bouncing around inside a 1.0 m × 1.0 m × 1.0 m box. The diameter of a Ping-Pong ball is 3.0 cm. What is the mean free path between collisions? Give your answer in cm.

Section 18.2 Pressure in a Gas

8. | Eleven molecules have speeds 15, 16, 17, . . . , 25 m/s. Calculate (a) v_{avg}, and (b) v_{rms}.

9. ‖ The molecules in a six-particle gas have velocities

$\vec{v}_1 = (20\hat{i} + 30\hat{j})$ m/s $\vec{v}_4 = (60\hat{i} - 20\hat{j})$ m/s

$\vec{v}_2 = (-40\hat{i} + 70\hat{j})$ m/s $\vec{v}_5 = -50\hat{j}$ m/s

$\vec{v}_3 = (-80\hat{i} - 10\hat{j})$ m/s $\vec{v}_6 = (40\hat{i} - 20\hat{j})$ m/s

Calculate (a) $\vec{v}_{avg}$, (b) v_{avg}, and (c) v_{rms}.

10. | FIGURE EX18.10 is a histogram showing the speeds of the molecules in a very small gas. What are (a) the most probable speed, (b) the average speed, and (c) the rms speed?

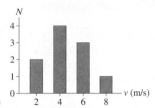

FIGURE EX18.10

11. ‖ The number density in a container of argon gas is 2.00×10^{25} m^{-3}. The atoms are moving with an rms speed of 455 m/s. What are (a) the pressure and (b) the temperature inside the container?

12. ‖ At 100°C the rms speed of nitrogen molecules is 576 m/s. Nitrogen at 100°C and a pressure of 2.0 atm is held in a container with a 10 cm × 10 cm square wall. Estimate the rate of molecular collisions (collisions/s) on this wall.

13. ‖ A cylinder contains gas at a pressure of 2.0 atm and a number density of 4.2×10^{25} m^{-3}. The rms speed of the atoms is 660 m/s. Identify the gas.

Section 18.3 Temperature

14. | What are the rms speeds of (a) neon atoms and (b) oxygen molecules at 1100°C?

15. | 1.5 m/s is a typical walking speed. At what temperature (in °C) would nitrogen molecules have an rms speed of 1.5 m/s?

16. | A gas consists of a mixture of neon and argon. The rms speed of the neon atoms is 400 m/s. What is the rms speed of the argon atoms?

17. ‖ At what temperature (in °C) do hydrogen molecules have the same rms speed as nitrogen molecules at 100°C?

18. | At what temperature (in °C) is the rms speed of oxygen molecules (a) half and (b) twice its value at STP?

19. | The rms speed of molecules in a gas is 400 m/s. What will be the rms speed if the gas pressure and volume are both doubled?

20. ‖ By what factor does the rms speed of a molecule change if the temperature is increased from 20°C to 100°C?

21. ‖ At what temperature would the rms speed of hydrogen molecules be the speed of light (3.0×10^8 m/s)? There is no upper limit to temperature, but Einstein's theory of relativity says that no material particle can attain the speed of light. Consequently, our results for ϵ_{avg} and v_{rms} would need to be modified for very high temperatures and speeds.

22. | Suppose you double the temperature of a gas at constant volume. Do the following change? If so, by what factor?
  a. The average translational kinetic energy of a molecule.
  b. The rms speed of a molecule.
  c. The mean free path.

23. | At STP, what is the total translational kinetic energy of the molecules in 1.0 mol of (a) hydrogen, (b) helium, and (c) oxygen?

24. ‖ During a physics experiment, helium gas is cooled to a temperature of 10 K at a pressure of 0.10 atm. What are (a) the mean free path in the gas, (b) the rms speed of the atoms, and (c) the average energy per atom?

25. | What are (a) the average kinetic energy and (b) the rms speed of a proton in the center of the sun, where the temperature is 2.0×10^7 K?

26. | The atmosphere of the sun consists mostly of hydrogen *atoms* (not molecules) at a temperature of 6000 K. What are (a) the average translational kinetic energy per atom and (b) the rms speed of the atoms?

Section 18.4 Thermal Energy and Specific Heat

27. | The average speed of the atoms in a 2.0 g sample of helium gas is 700 m/s. Estimate the thermal energy of the sample.

28. | A 10 g sample of neon gas has 1700 J of thermal energy. Estimate the average speed of a neon atom.

29. ‖ A 6.0 m × 8.0 m × 3.0 m room contains air at 20°C. What is the room's thermal energy?

30. ‖ What is the thermal energy of 100 cm^3 of lead at room temperature (20°C)?

31. | The thermal energy of 1.0 mol of a substance is increased by 1.0 J. What is the temperature change if the system is (a) a monatomic gas, (b) a diatomic gas, and (c) a solid?

32. | 1.0 mol of a monatomic gas interacts thermally with 1.0 mol of an elemental solid. The gas temperature decreases by 50°C at constant volume. What is the temperature change of the solid?

33. | A rigid container holds 0.20 g of hydrogen gas. How much heat is needed to change the temperature of the gas
  a. From 50 K to 100 K?
  b. From 250 K to 300 K?
  c. From 550 K to 600 K?
  d. From 2250 K to 2300 K?

34. | A cylinder of nitrogen gas has a volume of 15,000 cm^3 and a pressure of 100 atm.
  a. What is the thermal energy of this gas at room temperature (20°C)?
  b. What is the mean free path in the gas?
  c. The valve is opened and the gas is allowed to expand slowly and isothermally until it reaches a pressure of 1.0 atm. What is the change in the thermal energy of the gas?

Section 18.5 Thermal Interactions and Heat

35. | 2.0 mol of monatomic gas A initially has 5000 J of thermal energy. It interacts with 3.0 mol of monatomic gas B, which initially has 8000 J of thermal energy.
  a. Which gas has the higher initial temperature?
  b. What are the final thermal energies of each gas?

36. | 4.0 mol of monatomic gas A initially has 9000 J of thermal energy. It interacts with 3.0 mol of monatomic gas B, which initially has 5000 J of thermal energy. How much heat energy is transferred between the systems, and in which direction, as they come to thermal equilibrium?

Problems

37. ‖ For a monatomic gas, what is the ratio of the volume per atom (V/N) to the volume *of* an atom when the mean free path is ten times the atomic diameter?

38. ‖ From what height must an oxygen molecule fall in a vacuum so that its kinetic energy at the bottom equals the average energy of an oxygen molecule at 300 K?

39. ‖‖ A gas at $p = 50$ kPa and $T = 300$ K has a mass density of 0.0802 kg/m^3.
    a. Identify the gas.
    b. What is the rms speed of the atoms in this gas?
    c. What is the mean free path of the atoms in the gas?

40. ‖ Interstellar space, far from any stars, is filled with a very low density of hydrogen atoms (H, not H$_2$). The number density is about 1 atom/cm^3 and the temperature is about 3 K.
    a. Estimate the pressure in interstellar space. Give your answer in Pa and in atm.
    b. What is the rms speed of the atoms?
    c. What is the edge length L of an $L \times L \times L$ cube of gas with 1.0 J of thermal energy?

41. ‖ Dust particles are ≈ 10 μm in diameter. They are pulverized rock, with $\rho \approx 2500$ kg/m^3. If you treat dust as an ideal gas, what is the rms speed of a dust particle at 20°C?

42. ‖ Uranium has two naturally occurring isotopes. ^{238}U has a natural abundance of 99.3% and ^{235}U has an abundance of 0.7%. It is the rarer ^{235}U that is needed for nuclear reactors. The isotopes are separated by forming uranium hexafluoride UF$_6$, which is a gas, then allowing it to diffuse through a series of porous membranes. ^{235}UF$_6$ has a slightly larger rms speed than ^{238}UF$_6$ and diffuses slightly faster. Many repetitions of this procedure gradually separate the two isotopes. What is the ratio of the rms speed of ^{235}UF$_6$ to that of ^{238}UF$_6$?

43. ‖ Equation 18.3 is the mean free path of a particle through a gas of identical particles of equal radius. An electron can be thought of as a point particle with zero radius.
    a. Find an expression for the mean free path of an electron through a gas.
    b. Electrons travel 3 km through the Stanford Linear Accelerator (SLAC). In order for scattering losses to be negligible, the pressure inside the accelerator tube must be reduced to the point where the mean free path is at least 50 km. What is the maximum possible pressure inside the accelerator tube, assuming $T = 20$°C? Give your answer in both Pa and atm.

44. ‖‖‖ 5.0×10^{23} nitrogen molecules collide with a 10 cm^2 wall each second. Assume that the molecules all travel with a speed of 400 m/s and strike the wall head-on. What is the pressure on the wall?

45. ‖ A 10-cm-diameter, 20-cm-long cylinder contains 2.0×10^{22} atoms of argon at a temperature of 50°C.
    a. What is the number density of the gas?
    b. What is the root-mean-square speed?
    c. What is $(v_x)_{rms}$, the rms value of the x-component of velocity?
    d. What is the rate at which atoms collide with one end of the cylinder?
    e. Determine the pressure in the cylinder using the results of kinetic theory.
    f. Determine the pressure in the cylinder using the ideal-gas law.

46. ‖ A 10 cm $\times$ 10 cm $\times$ 10 cm box contains 0.010 mol of nitrogen at 20°C. What is the rate of collisions (collisions/s) on one wall of the box?

47. ‖ **FIGURE P18.47** shows the thermal energy of 0.14 mol of gas as a function of temperature. What is C_V for this gas?

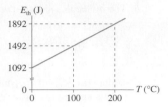

FIGURE P18.47

48. ‖ A 100 cm^3 box contains helium at a pressure of 2.0 atm and a temperature of 100°C. It is placed in thermal contact with a 200 cm^3 box containing argon at a pressure of 4.0 atm and a temperature of 400°C.
    a. What is the initial thermal energy of each gas?
    b. What is the final thermal energy of each gas?
    c. How much heat energy is transferred, and in which direction?
    d. What is the final temperature?
    e. What is the final pressure in each box?

49. ‖ 2.0 g of helium at an initial temperature of 300 K interacts thermally with 8.0 g of oxygen at an initial temperature of 600 K.
    a. What is the initial thermal energy of each gas?
    b. What is the final thermal energy of each gas?
    c. How much heat energy is transferred, and in which direction?
    d. What is the final temperature?

50. ‖ A gas of 1.0×10^{20} atoms or molecules has 1.0 J of thermal energy. Its molar specific heat at constant pressure is 20.8 J/mol K. What is the temperature of the gas?

51. ‖ How many degrees of freedom does a system have if $\gamma = 1.29$?

52. ‖ 1.0 mol of a monatomic gas and 1.0 mol of a diatomic gas are at 0°C. Both are heated at constant pressure until their volume doubles. What is the ratio $Q_{diatomic}/Q_{monatomic}$?

53. ‖ In the discussion following Equation 18.44 it was said that $Q_1 = -Q_2$. Prove that this is so.

54. ‖ A monatomic gas is adiabatically compressed to $\frac{1}{8}$ of its initial volume. Does each of the following quantities change? If so, does it increase or decrease, and by what factor? If not, why not?
    a. The rms speed.
    b. The mean free path.
    c. The thermal energy of the gas.
    d. The molar specific heat at constant volume.

55. ‖ Laser techniques can be used to confine a dilute gas of cesium atoms in a plane, forming a two-dimensional gas. What is the molar specific heat at (a) constant volume and (b) constant pressure for this gas? Give your answers as multiples of R.

56. ‖ Predict the molar specific heat at constant volume of (a) a two-dimensional monatomic gas and (b) a two-dimensional solid. Give your answers as multiples of R.

57. ‖ Equal masses of hydrogen gas and oxygen gas are mixed together in a container and held at constant temperature. What is the hydrogen/oxygen ratio of (a) v_{rms}, (b) ϵ_{avg}, and (c) E_{th}?

58. ‖ The rms speed of the molecules in 1.0 g of hydrogen gas is 1800 m/s.
    a. What is the total translational kinetic energy of the gas molecules?
    b. What is the thermal energy of the gas?
    c. 500 J of work are done to compress the gas while, in the same process, 1200 J of heat energy are transferred from the gas to the environment. Afterward, what is the rms speed of the molecules?

59. ‖ At what temperature does the rms speed of (a) a nitrogen molecule and (b) a hydrogen molecule equal the escape speed from the earth's surface? (c) You'll find that these temperatures are very high, so you might think that the earth's gravity could easily contain both gases. But not all molecules move with v_{rms}. There is a distribution of speeds, and a small percentage of molecules have speeds several times v_{rms}. Bit by bit, a gas can slowly leak out of the atmosphere as its fastest molecules escape. A reasonable rule of thumb is that the earth's gravity can contain a gas only if the average translational kinetic energy per molecule is less than 1% of the kinetic energy needed to escape. Use this rule to show why the earth's atmosphere contains nitrogen but not hydrogen, even though hydrogen is the most abundant element in the universe.

60. ‖ n_1 moles of a monatomic gas and n_2 moles of a diatomic gas are mixed together in a container.
    a. Derive an expression for the molar specific heat at constant volume of the mixture.
    b. Show that your expression has the expected behavior if $n_1 \to 0$ or $n_2 \to 0$.

61. ‖ A 1.0 kg ball is at rest on the floor in a 2.0 m × 2.0 m × 2.0 m room of air at STP. Air is 80% nitrogen (N_2) and 20% oxygen (O_2) by volume.
    a. What is the thermal energy of the air in the room?
    b. What fraction of the thermal energy would have to be conveyed to the ball for it to be spontaneously launched to a height of 1.0 m?
    c. By how much would the air temperature have to decrease to launch the ball?
    d. Your answer to part c is so small as to be unnoticeable, yet this event never happens. Why not?

62. ‖ An inventor wants you to invest money with his company, offering you 10% of all future profits. He reminds you that the brakes on cars get extremely hot when they stop and that there is a large quantity of thermal energy in the brakes. He has invented a device, he tells you, that converts that thermal energy into the forward motion of the car. This device will take over from the engine after a stop and accelerate the car back up to its original speed, thereby saving a tremendous amount of gasoline. Now, you're a smart person, so he admits up front that this device is not 100% efficient, that there is some unavoidable heat loss to the air and to friction within the device, but the upcoming research for which he needs your investment will make those losses extremely small. You do also have to start the car with cold brakes after it has been parked awhile, so you'll still need a gasoline engine for that. Nonetheless, he tells you, his prototype car gets 500 miles to the gallon and he expects to be at well over 1000 miles to the gallon after the next phase of research. Should you invest? Base your answer on an analysis of the *physics* of the situation.

Challenge Problems

63. 1.0 mol of a diatomic gas with $C_V = \frac{5}{2}R$ has initial pressure p_i and volume V_i. The gas undergoes a process in which the pressure is directly proportional to the volume until the rms speed of the molecules has doubled.
    a. Show this process on a pV diagram.
    b. How much heat does this process require? Give your answer in terms of p_i and V_i.

64. An experiment you're designing needs a gas with $\gamma = 1.50$. You recall from your physics class that no individual gas has this value, but it occurs to you that you could produce a gas with $\gamma = 1.50$ by mixing together a monatomic gas and a diatomic gas. What fraction of the molecules need to be monatomic?

65. Consider a container like that shown in Figure 18.14, with n_1 moles of a monatomic gas on one side and n_2 moles of a diatomic gas on the other. The monatomic gas has initial temperature T_{1i}. The diatomic gas has initial temperature T_{2i}.
    a. Show that the equilibrium thermal energies are
    $$E_{1f} = \frac{3n_1}{3n_1 + 5n_2}(E_{1i} + E_{2i})$$
    $$E_{2f} = \frac{5n_2}{3n_1 + 5n_2}(E_{1i} + E_{2i})$$
    b. Show that the equilibrium temperature is
    $$T_f = \frac{3n_1 T_{1i} + 5n_2 T_{2i}}{3n_1 + 5n_2}$$
    c. 2.0 g of helium at an initial temperature of 300 K interacts thermally with 8.0 g of oxygen at an initial temperature of 600 K. What is the final temperature? How much heat energy is transferred, and in which direction?

STOP TO THINK ANSWERS

Stop to Think 18.1: $\lambda_B > \lambda_A = \lambda_C > \lambda_D$. Increasing the volume makes the gas less dense, so λ increases. Increasing the radius makes the targets larger, so λ decreases. The mean free path doesn't depend on the atomic mass.

Stop to Think 18.2: c. Each v^2 increases by a factor of 16 but, after averaging, v_{rms} takes the square root.

Stop to Think 18.3: c. The average translational kinetic energy per molecule depends *only* on the temperature.

Stop to Think 18.4: b. The bead can slide along the wire (one degree of translational motion) and rotate around the wire (one degree of rotational motion).

Stop to Think 18.5: a. Temperature measures the average translational kinetic energy *per molecule,* not the thermal energy of the entire system.

Stop to Think 18.6: c. With 1,000,000 molecules, it's highly unlikely that 750,000 of them would spontaneously move into one side of the box. A state with a very small probability of occurrence has a very low entropy. Having an imbalance of only 100 out of 1,000,000 is well within what you might expect for random fluctuations. This is a highly probable situation and thus one of large entropy.

19 Heat Engines and Refrigerators

That's not smoke. It's clouds of water vapor rising from the cooling towers around a large power plant. The power plant is generating electricity by turning heat into work—but not very efficiently. Roughly two-thirds of the fuel's energy is being dissipated into the air as "waste heat."

▶ **Looking Ahead**

The goal of Chapter 19 is to study the physical principles that govern the operation of heat engines and refrigerators. In this chapter you will learn to:

- Understand and analyze heat engines and refrigerators.
- Understand the concept and significance of the Carnot engine.
- Characterize the performance of a heat engine in terms of its thermal efficiency and that of a refrigerator in terms of its coefficient of performance.
- Recognize that the second law of thermodynamics limits the efficiencies of heat engines.

◀ **Looking Back**

The material in this chapter depends on the first and second laws of thermodynamics. Most of the examples will be based on ideal gases. Please review:

- Sections 16.5–16.6 Ideal gases.
- Sections 17.2–17.4 Work, heat, and the first law of thermodynamics.
- Section 18.6 The second law of thermodynamics.

The earliest humans learned to use the heat from fires to warm themselves and cook their food. They were transforming heat energy into thermal energy. But is there a way to transform heat into *work?* Can we use the energy released by the fuel to grind corn, pump water, accelerate cars, launch rockets, or do any other task in which a force is exerted through a distance?

The first practical device for turning heat into work was the steam engine, the symbol of the Industrial Revolution. A steam engine boils water to make high-pressure steam, then uses the steam to push a piston and do work. The 19th and 20th centuries saw the development of the steam turbine, the gasoline engine, the jet engine, and other devices that transform the heat from burning fuel into useful work. These are the devices that power modern society.

"Heat engine" is the generic term for *any* device that uses a cyclical process to transform heat energy into work. The power plant shown in the photo and the engine in your car are examples of heat engines. A closely related concept is a *refrigerator,* a device that uses work to move heat energy from a cold object to a hot object. Our goal in this chapter is to investigate the physical principles that *all* heat engines and *all* refrigerators must obey. We'll discover that the second law of thermodynamics places sharp constraints on the maximum possible efficiency of heat engines and refrigerators.

19.1 Turning Heat into Work

Thermodynamics is the branch of physics that studies the transformation of energy. Many practical devices are designed to transform energy from one form, such as the heat from burning fuel, into another, such as work. Chapters 17 and 18 established two laws of thermodynamics that any such device must obey:

> **First law** Energy is conserved; that is, $\Delta E_{th} = W + Q$.
>
> **Second law** Most macroscopic processes are irreversible. In particular, heat energy is transferred spontaneously from a hotter to a colder system but never from a colder system to a hotter system.

A car engine transforms the chemical energy stored in the fuel into work and ultimately into the car's kinetic energy.

Our goal in this chapter is to discover what these two laws, especially the second law, imply about devices that turn heat into work. In particular:

- How does a practical device transform heat into work?
- What are the limitations and restrictions on these energy transformations?

Much of this chapter will be an exercise in logical deduction. The reasoning is subtle but important.

Work Done by the System

In mechanics, "work" means the work done *on* the system by an external force. However, it is useful in practical thermodynamics to turn things around and speak of the work done on the environment *by* the system.

In **FIGURE 19.1a**, the gas pressure pushes outward on the piston with force $\vec{F}_{gas}$. Some object in the environment, usually a piston rod, pushes inward with force $\vec{F}_{ext}$. This external force keeps the gas pressure from blowing the piston out. For any quasi-static process, where the system is essentially in equilibrium at all times, these two forces must balance: $\vec{F}_{gas} = -\vec{F}_{ext}$.

The work W done *on* the system is the work done by $\vec{F}_{ext}$ as the piston moves through a displacement Δx. You learned in Chapter 17 that W is the *negative* of the area under the pV curve of the process. But force $\vec{F}_{gas}$ also does work on the moving piston. Because $\vec{F}_{gas} = -\vec{F}_{ext}$, the work done by $\vec{F}_{gas}$, which we call the work W_s done *by* the system, has the same absolute value as the work W but the opposite sign. As **FIGURE 19.1b** shows, the work done *by* the system is

FIGURE 19.1 Forces $\vec{F}_{gas}$ and $\vec{F}_{ext}$ both do work as the piston moves.

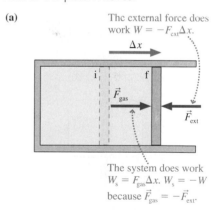

(a)

The external force does work $W = -F_{ext}\Delta x$.

The system does work $W_s = F_{gas}\Delta x$. $W_s = -W$ because $\vec{F}_{gas} = -\vec{F}_{ext}$.

$$W_s = -W = \text{the area under the } pV \text{ curve} \qquad (19.1)$$

W_s is positive when energy is transferred *out* of the system.

Work done by the environment and work done by the system are not mutually exclusive. Both $\vec{F}_{gas}$ and $\vec{F}_{ext}$ do work as the piston moves. Energy is transferred *into* the system as a gas is compressed; hence W is positive and W_s is negative. Energy is transferred *out* of the system as a gas expands; thus W is negative and W_s is positive.

> **NOTE** ▶ When energy is transferred *into* a system, by compressing the gas, it is customary to say "the environment does work on the system." Similarly, when the gas pushes the piston out and transfers energy *out* of the system, we customarily say "the system does work on the environment." Neither is meant to imply that the "other" work isn't being done at the same time. ◀

The first law of thermodynamics $\Delta E_{th} = W + Q$ can be written in terms of W_s as

$$Q = W_s + \Delta E_{th} \qquad \text{(first law of thermodynamics)} \qquad (19.2)$$

It's easy to interpret this version of the first law. Because energy must be conserved, **any energy transferred into a system as heat is either used to do work or stored within the system as an increased thermal energy.**

(b)

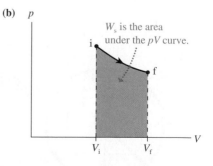

W_s is the area under the pV curve.

Energy-Transfer Diagrams

Suppose you drop a hot rock into the ocean. Heat is transferred from the rock to the ocean until the rock and ocean are at the same temperature. Although the ocean warms up ever so slightly, ΔT_{ocean} is so small as to be completely insignificant. For all practical purposes, the ocean is infinite and unchangeable.

An **energy reservoir** is an object or a part of the environment so large that its temperature does not change when heat is transferred between the system and the reservoir. A reservoir at a higher temperature than the system is called a *hot reservoir*. A vigorously burning flame is a hot reservoir for small objects placed in the flame. A reservoir at a lower temperature than the system is called a *cold reservoir*. The ocean is a cold reservoir for the hot rock. We will use T_H and T_C to designate the temperatures of the hot and cold reservoirs.

Hot and cold reservoirs are idealizations, in the same category as frictionless surfaces and massless strings. No real object can maintain a perfectly constant temperature as heat is transferred in or out. Even so, an object can be modeled as a reservoir if it is much larger than the system that thermally interacts with it.

Heat energy is transferred between a system and a reservoir if they have different temperatures. We will define

Q_H = amount of heat transferred to or from a hot reservoir

Q_C = amount of heat transferred to or from a cold reservoir

By definition, Q_H and Q_C are *positive* quantities. The direction of heat transfer, which determines the sign of Q in the first law, will always be clear as we deal with thermodynamic devices. For example, the heat transferred *from* the system to a cold reservoir is $Q = -Q_C$.

FIGURE 19.2a shows a heavy copper bar between a hot reservoir (at temperature T_H) and a cold reservoir (at temperature T_C). Heat Q_H is transferred from the hot reservoir into the copper and heat Q_C is transferred from the copper to the cold reservoir. **FIGURE 19.2b** is an **energy-transfer diagram** for this process. The hot reservoir is always drawn at the top, the cold reservoir at the bottom, and the system—the copper bar in this case—between them. The reservoirs and the system are connected by "pipes" that show the energy transfers. Figure 19.2b shows heat Q_H being transferred into the system and Q_C being transferred out.

The first law of thermodynamics $Q = W_s + \Delta E_{\text{th}}$ refers to the *system*. Q is the net heat to the system. In this case, because Q_C is the quantity of heat *leaving* the system, $Q = Q_H - Q_C$. The copper bar does no work, so $W_s = 0$. The bar warms up when first placed between the two reservoirs, but it soon comes to a steady state where its temperature no longer changes. Then $\Delta E_{\text{th}} = 0$. Thus the first law tells us that $Q = Q_H - Q_C = 0$, from which we conclude that

$$Q_C = Q_H \tag{19.3}$$

In other words, all of the heat transferred into the hot end of the rod is subsequently transferred out of the cold end. This isn't surprising. After all, we know that heat is transferred spontaneously from a hotter object to a colder object. Even so, there has to be some *means* by which the heat energy gets from the hotter object to the colder. The copper bar provides a route for the energy transfer, and $Q_C = Q_H$ is the statement that energy is conserved as it moves through the bar.

Contrast Figure 19.2b with **FIGURE 19.2c**. Figure 19.2c shows a system in which heat is being transferred from the cold reservoir to the hot reservoir. The first law of thermodynamics is not violated, because $Q_H = Q_C$, but the second law is. If there were such a system, it would allow the spontaneous (i.e., with no outside input or assistance) transfer of heat from a colder object to a hotter object. The process of Figure 19.2c is forbidden by the second law of thermodynamics.

FIGURE 19.2 Energy-transfer diagrams.

(a)

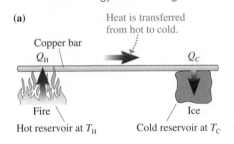

Heat is transferred from hot to cold.

Copper bar

Q_H Q_C

Fire Ice

Hot reservoir at T_H Cold reservoir at T_C

(b)

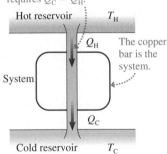

Heat energy is transferred from a hot reservoir to a cold reservoir. Energy conservation requires $Q_C = Q_H$.

Hot reservoir T_H

Q_H The copper bar is the system.

System

Q_C

Cold reservoir T_C

(c)

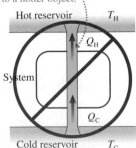

The second law forbids a process in which heat is spontaneously transferred from a colder object to a hotter object.

Hot reservoir T_H

Q_H

System

Q_C

Cold reservoir T_C

Work into Heat and Heat into Work

Turning work into heat is easy. Take two rocks out of the ocean and rub them together vigorously until both are warmer. This is a mechanical interaction in which work increases the thermal energy of the rocks, or $W \rightarrow \Delta E_{th}$. Then toss both back into the ocean, where they return to their initial temperature as thermal energy is transferred as heat from the slightly warmer rocks to the colder water ($\Delta E_{th} \rightarrow Q_C$). **FIGURE 19.3** is the energy-transfer diagram for this process.

NOTE ▶ Energy-transfer diagrams show the "work pipe" entering or leaving the system from the side. ◀

The conversion of work into heat is 100% efficient. That is, *all* of the energy supplied to the system as work W is transferred into the ocean as heat Q_C. This perfect transformation of work into heat can continue as long as there is motion. (It was this continual production of heat energy in the boring of cannons that Count Rumford recognized as being in conflict with the caloric theory.)

But the reverse—transforming heat into work—isn't so easy.

FIGURE 19.4 shows an isothermal process in which the temperature remains constant because the heat energy from the flame is used to do the work of lifting the mass. $\Delta E_{th} = 0$ in an isothermal expansion, so the first law is

$$W_s = Q \tag{19.4}$$

The energy that is transferred into the gas as heat is transformed with 100% efficiency into work done by the gas as it lifts the mass. So why did we just say that transforming heat into work isn't as easy as transforming work into heat?

There's a difference. In Figure 19.3, where we transformed work into heat, the system *returned to its initial state.* We can repeat the process over and over, continuing to transform work into heat as long as there is motion. But Figure 19.4 is a one-time process. The gas does work once as it lifts the piston, but then the gas is no longer in its initial state. We cannot repeat the process. Extracting more and more work from the device of Figure 19.4 requires lifting the piston higher and higher until, ultimately, it reaches the end of the cylinder.

To be practical, **a device that transforms heat into work must return to its initial state at the end of the process and be ready for continued use.** You want your car engine to turn over and over as long as there is fuel.

Perhaps Figure 19.4 is just a bad idea for turning heat into work. Perhaps some other device can turn heat into work continuously. Interestingly, no one has ever invented a "perfect engine" that transforms heat into work with 100% efficiency *and returns to its initial state* so that it can continue to do work as long as there is fuel. Of course, that such a device has not been invented is not a proof that it can't be done. We'll provide a proof shortly, but for now we'll make the hypothesis that the process of **FIGURE 19.5** is somehow forbidden.

Notice the asymmetry between Figures 19.3 and 19.5. The perfect transformation of work into heat is permitted, but the perfect transformation of heat into work is forbidden. This asymmetry parallels the asymmetry of the two processes in Figure 19.2. In fact, we'll soon see that the "perfect engine" of Figure 19.5 is forbidden for exactly the same reason: the second law of thermodynamics.

19.2 Heat Engines and Refrigerators

The steam generator at your local electric power plant works by boiling water to produce high-pressure steam that spins a turbine (which then spins a generator to produce electricity). That is, the steam pressure is doing work. The steam is then condensed to liquid water and pumped back to the boiler to start the process again. There are two crucial ideas here. First, the device works in a cycle, with the water returning to its

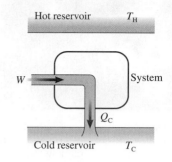

FIGURE 19.3 Work can be transformed into heat with 100% efficiency.

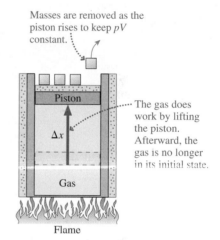

FIGURE 19.4 An isothermal process transforms heat into work, but only as a one-time event.

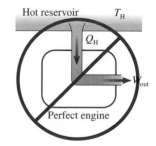

FIGURE 19.5 There are no perfect engines that turn heat into work with 100% efficiency.

The steam turbine in a modern power plant is an enormous device. Expanding steam does work by spinning the turbine.

initial conditions once a cycle. Second, heat is transferred to the water in the boiler, but heat is transferred *out* of the water in the condenser.

Car engines and steam generators are examples of what we call *heat engines*. A **heat engine** is any closed-cycle device that extracts heat Q_H from a hot reservoir, does useful work, and exhausts heat Q_C to a cold reservoir. A **closed-cycle device** is one that periodically *returns to its initial conditions,* repeating the same process over and over. That is, all state variables (pressure, temperature, thermal energy, and so on) return to their initial values once every cycle. Consequently, a heat engine can continue to do useful work for as long as it is attached to the reservoirs.

FIGURE 19.6 is the energy-transfer diagram of a heat engine. Unlike the forbidden "perfect engine" of Figure 19.5, a heat engine is connected to both a hot reservoir *and* a cold reservoir. You can think of a heat engine as "siphoning off" some of the heat that moves from the hot reservoir to the cold reservoir and transforming that heat into work—some heat, but not all.

Because the temperature and thermal energy of a heat engine return to their initial values at the end of each cycle, there is no *net* change in E_{th}:

$$(\Delta E_{th})_{net} = 0 \qquad \text{(any heat engine, over one full cycle)} \qquad (19.5)$$

Consequently, the first law of thermodynamics *for a full cycle* of a heat engine is $(\Delta E_{th})_{net} = Q - W_s = 0$.

Let's define W_{out} to be the useful work done *by* the heat engine *per cycle*. The net heat transfer per cycle is $Q_{net} = Q_H - Q_C$; hence the first law applied to a heat engine is

$$W_{out} = Q_{net} = Q_H - Q_C \quad \text{(work per cycle done by a heat engine)} \qquad (19.6)$$

This is just energy conservation. The energy transferred into the engine (Q_H) and energy transferred out of the engine (Q_C and W_{out}) have to balance. The energy-transfer diagram of Figure 19.6 is a pictorial representation of Equation 19.6.

FIGURE 19.6 The energy-transfer diagram of a heat engine.

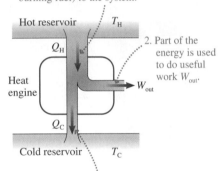

1. Heat energy Q_H is transferred from the hot reservoir (typically burning fuel) to the system.

2. Part of the energy is used to do useful work W_{out}.

3. The remaining energy $Q_C = Q_H - W_{out}$ is exhausted to the cold reservoir (cooling water or the air) as waste heat.

NOTE ▶ Equations 19.5 and 19.6 apply only to a *full cycle* of the heat engine. They are *not* valid for any of the individual processes that make up a cycle. ◀

For practical reasons, we would like an engine to do the maximum amount of work with the minimum amount of fuel. We can measure the performance of a heat engine in terms of its **thermal efficiency** η (lowercase Greek eta), defined as

$$\eta = \frac{W_{out}}{Q_H} = \frac{\text{what you get}}{\text{what you had to pay}} \qquad (19.7)$$

Using Equation 19.6 for W_{out}, we can also write the thermal efficiency as

$$\eta = 1 - \frac{Q_C}{Q_H} \qquad (19.8)$$

FIGURE 19.7 η is the fraction of heat energy that is transformed into useful work.

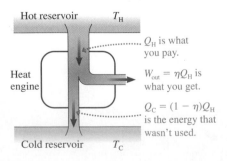

Hot reservoir T_H

Q_H is what you pay.

$W_{out} = \eta Q_H$ is what you get.

$Q_C = (1 - \eta)Q_H$ is the energy that wasn't used.

Cold reservoir T_C

FIGURE 19.7 illustrates the idea of thermal efficiency.

A *perfect* heat engine would have $\eta_{perfect} = 1$. That is, it would be 100% efficient at converting heat from the hot reservoir (the burning fuel) into work. You can see from Equation 19.8 that a perfect engine would have no exhaust ($Q_C = 0$) and would not need a cold reservoir. Figure 19.5 has already suggested that there are no perfect heat engines, that an engine with $\eta = 1$ is impossible. A heat engine *must* exhaust **waste heat** to a cold reservoir. It is energy that was extracted from the hot reservoir but *not* transformed to useful work.

Practical heat engines, such as car engines and steam generators, have thermal efficiencies in the range $\eta \approx 0.1$–0.5. This is not large. Can a clever designer do better, or is this some kind of physical limitation?

Rank in order, from largest to smallest, the work W_{out} performed by these four heat engines.

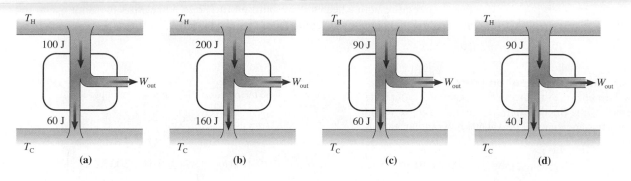

A Heat-Engine Example

To illustrate how these ideas actually work, **FIGURE 19.8** shows a simple engine that converts heat into the work of lifting mass M. The gas does work on the environment while lifting the mass during step (b) (W_s is positive, W is negative). A steadily increasing force from the environment, perhaps due to a piston rod, does work on the gas during the compression of step (e) (W is positive, W_s is negative).

FIGURE 19.8 A simple heat engine transforms heat into work.

(a) Heat is transferred into the gas from the burning fuel.

(b) The gas does work by lifting the mass in an isobaric expansion.

(c) The piston is locked and the mass is removed. The heat is turned off.

(d) The gas cools back to room temperature at constant volume. Then the piston is unlocked.

(e) A steadily increasing external force steadily raises the pressure in an isothermal compression until the pressure has been restored to its initial value.

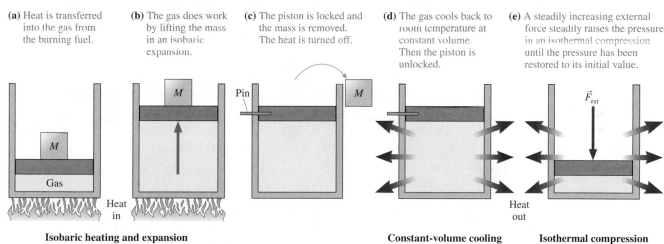

Isobaric heating and expansion   **Constant-volume cooling**   **Isothermal compression**

The net effect of this multistep process is to convert some of the fuel's energy into the useful work of lifting the mass. There has been no net change in the gas, which has returned to its initial pressure, volume, and temperature at the end of step (e). We can start the whole process over again and continue lifting masses (doing work) as long as we have fuel.

FIGURE 19.9 shows the heat-engine process on a pV diagram. It is a *closed cycle* because the gas returns to its initial conditions. No work is done during the isochoric process, and, as you can see from the areas under the curve, the work done *by* the gas to lift the mass is greater than the work the environment must do *on* the gas to recompress it. Thus this heat engine, by burning fuel, does *net* work per cycle: $W_{net} = W_{lift} - W_{ext} = (W_s)_{1\rightarrow 2} + (W_s)_{3\rightarrow 1}$.

Notice that the cyclical process of Figure 19.9 involves two *cooling processes* in which heat is transferred *from* the gas to the environment. Heat energy is transferred

FIGURE 19.9 The closed-cycle pV diagram for the heat engine of Figure 19.8.

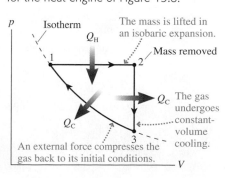

from hotter objects to colder objects, so the system *must* be connected to a cold reservoir with $T_C < T_{gas}$ during these two processes. A key to understanding heat engines is that they require both a heat source (burning fuel) *and* a heat sink (cooling water, the air, or something at a lower temperature than the system).

EXAMPLE 19.1 Analyzing a heat engine I

Analyze the heat engine of FIGURE 19.10 to determine (a) the net work done per cycle, (b) the engine's thermal efficiency, and (c) the engine's power output if it runs at 600 rpm. Assume the gas is monatomic.

FIGURE 19.10 The heat engine of Example 19.1.

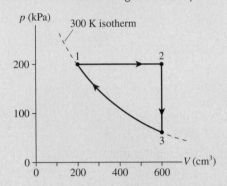

MODEL The gas follows a closed cycle consisting of three distinct processes, each of which was studied in Chapters 16 and 17. For each of the three we need to determine the work done and the heat transferred.

SOLVE To begin, we can use the initial conditions at state 1 and the ideal-gas law to determine the number of moles of gas:

$$n = \frac{p_1 V_1}{RT_1} = \frac{(200 \times 10^3 \text{ Pa})(2.0 \times 10^{-4} \text{ m}^3)}{(8.31 \text{ J/mol K})(300 \text{ K})} = 0.0160 \text{ mol}$$

Process 1 → 2: The work done *by* the gas in the isobaric expansion is

$$(W_s)_{12} = p\Delta V = (200 \times 10^3 \text{ Pa})(6.0 - 2.0) \times 10^{-4} \text{ m}^3$$
$$= 80 \text{ J}$$

We can use the ideal-gas law at constant pressure to find $T_2 = (V_2/V_1)T_1 = 3T_1 = 900$ K. The heat transfer during a constant-pressure process is

$$Q_{12} = nC_P\Delta T$$
$$= (0.0160 \text{ mol})(20.8 \text{ J/mol K})(900 \text{ K} - 300 \text{ K})$$
$$= 200 \text{ J}$$

where we used $C_P = \frac{5}{2}R$ for a monatomic ideal gas.

Process 2 → 3: No work is done in an isochoric process, so $(W_s)_{23} = 0$. The temperature drops back to 300 K, so the heat transfer is

$$Q_{23} = nC_V\Delta T$$
$$= (0.0160 \text{ mol})(12.5 \text{ J/mol K})(300 \text{ K} - 900 \text{ K})$$
$$= -120 \text{ J}$$

where we used $C_V = \frac{3}{2}R$.

Process 3 → 1: The gas returns to its initial state with volume V_1. The work done *by* the gas during an isothermal process is

$$(W_s)_{31} = nRT \ln\left(\frac{V_1}{V_3}\right)$$
$$= (0.0160 \text{ mol})(8.31 \text{ J/mol K})(300 \text{ K}) \ln\left(\frac{1}{3}\right)$$
$$= -44 \text{ J}$$

W_s is negative because the environment does work on the gas to compress it. An isothermal process has $\Delta E_{th} = 0$ and hence, from the first law,

$$Q_{31} = (W_s)_{31} = -44 \text{ J}$$

Q is negative because the gas must be cooled as it is compressed to keep the temperature constant.

a. The *net* work done by the engine during one cycle is

$$W_{out} = (W_s)_{12} + (W_s)_{23} + (W_s)_{31} = 36 \text{ J}$$

As a consistency check, notice that the net heat transfer is

$$Q_{net} = Q_{12} + Q_{23} + Q_{31} = 36 \text{ J}$$

Equation 19.6 told us that a heat engine *must* have $W_{out} = Q_{net}$, and we see that it does.

b. The efficiency depends not on the net heat transfer but on the heat Q_H transferred into the engine from the flame. Heat is transferred in during process 1 → 2, where Q is positive, and out during processes 2 → 3 and 3 → 1, where Q is negative. Thus

$$Q_H = Q_{12} = 200 \text{ J}$$
$$Q_C = |Q_{23}| + |Q_{31}| = 164 \text{ J}$$

Notice that $Q_H - Q_C = 36$ J $= W_{out}$. In this heat engine, 200 J of heat from the hot reservoir does 36 J of useful work. Thus the thermal efficiency is

$$\eta = \frac{W_{out}}{Q_H} = \frac{36 \text{ J}}{200 \text{ J}} = 0.18 \text{ or } 18\%$$

This heat engine is far from being a perfect engine!

c. An engine running at 600 rpm goes through 10 cycles per second. The power output is the work done *per second:*

$$P_{out} = (\text{work per cycle}) \times (\text{cycles per second})$$
$$= 360 \text{ J/s} = 360 \text{ W}$$

ASSESS Although we didn't need Q_{net}, verifying that $Q_{net} = W_{out}$ was a check of self-consistency. Heat-engine analysis requires many calculations and offers many opportunities to get signs wrong. However, there are a sufficient number of self-consistency checks so that you can almost always spot calculational errors *if you check for them.*

Let's think about this example a bit more before going on. We've said that a heat engine operates between a hot reservoir and a cold reservoir. Figure 19.10 doesn't explicitly show the reservoirs. Nonetheless, we know that heat is transferred from a hotter object to a colder object. Heat Q_H is transferred into the system during process $1 \rightarrow 2$ as the gas warms from 300 K to 900 K. For this to be true, the hot-reservoir temperature T_H must be $\geq$900 K. Likewise, heat Q_C is transferred from the system to the cold reservoir as the temperature drops from 900 K to 300 K in process $2 \rightarrow 3$. For this to be true, the cold-reservoir temperature T_C must be $\leq$300 K.

So we really don't know what the reservoirs are or their exact temperatures, but we can say with certainty that the hot-reservoir temperature T_H must exceed the highest temperature reached by the system and the cold-reservoir temperature T_C must be less than the coldest system temperature.

Refrigerators

Your house or apartment has a refrigerator. Very likely it has an air conditioner. The purpose of these devices is to make air that is cooler than its environment even colder. The first does so by blowing hot air out into a warm room, the second by blowing it out to the hot outdoors. You've probably felt the hot air exhausted by an air conditioner compressor or coming out from beneath the refrigerator.

At first glance, a refrigerator or air conditioner may seem to violate the second law of thermodynamics. After all, doesn't the second law forbid heat from being transferred from a colder object to a hotter object? Not quite: The second law says that heat is not *spontaneously* transferred from a colder to a hotter object. A refrigerator or air conditioner requires electric power to operate. They do cause heat to be transferred from cold to hot, but the transfer is "assisted" rather than spontaneous.

A **refrigerator** is any closed-cycle device that uses external work W_{in} to remove heat Q_C from a cold reservoir and exhaust heat Q_H to a hot reservoir. **FIGURE 19.11** is the energy-transfer diagram of a refrigerator. The cold reservoir is the air inside the refrigerator or the air inside your house on a summer day. To keep the air cold, in the face of inevitable "heat leaks," the refrigerator or air conditioner compressor continuously removes heat from the cold reservoir and exhausts heat into the room or outdoors. You can think of a refrigerator as "pumping heat uphill," much as a water pump lifts water uphill.

Because a refrigerator, like a heat engine, is a cyclical device, $\Delta E_{th} = 0$. Conservation of energy requires

$$Q_H = Q_C + W_{in} \qquad (19.9)$$

To move energy from a colder to a hotter reservoir, a refrigerator must exhaust *more* heat to the outside than it removes from the inside. This has significant implications for whether or not you can cool a room by leaving the refrigerator door open.

The thermal efficiency of a heat engine was defined as "what you get (useful work W_{out})" versus "what you had to pay (fuel to supply Q_H)." By analogy, we define the **coefficient of performance** K of a refrigerator to be

$$K = \frac{Q_C}{W_{in}} = \frac{\text{what you get}}{\text{what you had to pay}} \qquad (19.10)$$

What you get, in this case, is the removal of heat from the cold reservoir. But you have to pay the electric company for the work needed to run the refrigerator. A better refrigerator will require less work to remove a given amount of heat, thus having a larger coefficient of performance.

A perfect refrigerator would require no work ($W_{in} = 0$) and would have $K_{perfect} = \infty$. But if Figure 19.11 had no work input, it would look like Figure 19.2c. That device was forbidden by the second law of thermodynamics because, with no work input, heat would move *spontaneously* from cold to hot.

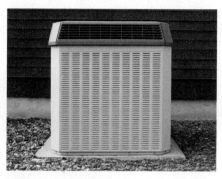

This air conditioner transfers heat energy *from* the cool indoors *to* the hot exterior.

FIGURE 19.11 The energy-transfer diagram of a refrigerator.

The amount of heat exhausted to the hot reservoir is larger than the amount of heat extracted from the cold reservoir.

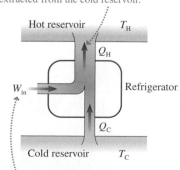

External work is used to remove heat from a cold reservoir and exhaust heat to a hot reservoir.

We noted in Chapter 18 that the second law of thermodynamics can be stated several different but equivalent ways. We can now give a third statement:

Second law, informal statement #3 There are no perfect refrigerators with coefficient of performance $K = \infty$.

Any real refrigerator or air conditioner *must* use work to move energy from the cold reservoir to the hot reservoir, hence $K < \infty$.

No Perfect Heat Engines

We hypothesized above that there are no perfect heat engines—that is, no heat engines like the one shown in Figure 19.5 with $Q_C = 0$ and $\eta = 1$. Now we're ready to prove this hypothesis. **FIGURE 19.12** shows a hot reservoir at temperature T_H and a cold reservoir at temperature T_C. An ordinary refrigerator, one that obeys all the laws of physics, is operating between these two reservoirs.

FIGURE 19.12 A perfect engine driving an ordinary refrigerator would be able to violate the second law of thermodynamics.

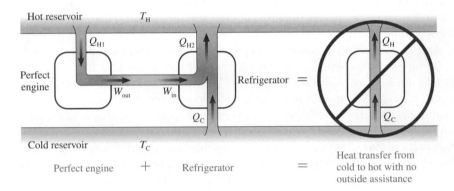

Suppose we had a perfect heat engine, one that takes in heat Q_H from the high-temperature reservoir and transforms that energy entirely into work W_{out}. If we had such a heat engine, we could use its output to provide the work input to the refrigerator. The two devices combined have no connection to the external world. That is, there's no net input or net output of work.

If we built a box around the heat engine and refrigerator, so that you couldn't see what was inside, the only thing you would observe is heat being transferred *with no outside assistance* from the cold reservoir to the hot reservoir. But a spontaneous or unassisted transfer of heat from a colder to a hotter object is exactly what the second law of thermodynamics forbids. Consequently, our assumption of a perfect heat engine must be wrong. Hence another statement of the second law of thermodynamics is:

Second law, informal statement #4 There are no perfect heat engines with efficiency $\eta = 1$.

Any real heat engine *must* exhaust waste heat Q_C to a cold reservoir.

Unanswered Questions

We noted that this chapter would be an exercise in logical deduction. By using only energy conservation and the fact that heat is not spontaneously transferred from cold to hot, we've been able to deduce that

- Heat engines and refrigerators exist.
- They must use a closed-cycle process, with $(\Delta E_{th})_{net} = 0$.
- There are no perfect heat engines. A heat engine *must* exhaust heat to a cold reservoir.
- There are no perfect refrigerators. A refrigerator *must* use external work.

This is a good start, but it leaves some unanswered questions. For example,

- With good design, can we make a heat engine whose thermal efficiency η approaches 1? Or is there an upper limit η_{max} that cannot be exceeded?
- If η has a maximum value, what is it?
- Likewise, is there an upper limit K_{max} for the coefficient of performance of a refrigerator? If so, what is it?

There is, indeed, an upper limit to η that no heat engine can exceed and an upper limit to K that no refrigerator can exceed. We'll be able to establish an actual value for η_{max} and find that, for many practical engines, η_{max} is distressingly low.

STOP TO THINK 19.2 It's a hot day and your air conditioner is broken. Your roommate says, "Let's open the refrigerator door and cool this place off." Will this work?

  a. Yes.   b. No.   c. It might, but it will depend on how hot the room is.

19.3 Ideal-Gas Heat Engines

We will focus on heat engines that use a gas as the *working substance*. The gasoline or diesel engine in your car is an engine that alternately compresses and expands a gaseous fuel-air mixture. Engines such as steam generators that rely on phase changes will be deferred to more advanced courses.

A gas heat engine can be represented by a closed-cycle trajectory in the pV diagram, such as the one shown in **FIGURE 19.13a**. This observation leads to an important geometric interpretation of the work done by the system during one full cycle. You learned in Section 19.1 that the work done *by* the system is the area under the curve of a pV trajectory. As **FIGURE 19.13b** shows, the work done during a full cycle is the work W_{expand} done by the system as it expands to V_{max} plus the work $W_{compress}$ done by the system as it is compressed back to V_{min}. That is,

$$W_{out} = W_{expand} - |W_{compress}| = \text{area } inside \text{ the closed curve} \quad (19.11)$$

FIGURE 19.13 The work W_{out} done by the system during one full cycle is the area enclosed within the curve.

(a)

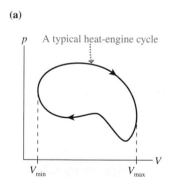

A typical heat-engine cycle

(b) As the gas expands, the work W_{expand} done by the gas is positive.

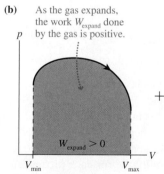

$W_{expand} > 0$

As the gas is compressed, the work $W_{compress}$ done by the gas is negative.

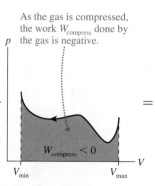

$W_{compress} < 0$

The net work done by the gas is the area enclosed within the curve.

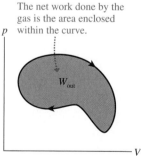

W_{out}

You can see that **the net work done by a gas heat engine during one full cycle is the area enclosed by the pV curve for the cycle.** A thermodynamic cycle with a larger enclosed area does more work than one with a smaller enclosed area. Notice that the gas must go around the pV trajectory in a *clockwise* direction for W_{out} to be positive. We'll see later that a refrigerator uses a counterclockwise (ccw) cycle.

Ideal-Gas Summary

We've learned a lot about ideal gases in the last three chapters. All gas processes obey the ideal-gas law $pV = nRT$ and the first law of thermodynamics $\Delta E_{th} = Q - W_s$.

Table 19.1 summarizes the results for specific gas processes. This table shows W_s, the work done *by* the system, so the signs are opposite those in Chapter 17.

TABLE 19.1 Summary of ideal-gas processes

Process	Gas law	Work W_s	Heat Q	Thermal energy
Isochoric	$p_i/T_i = p_f/T_f$	0	$nC_V\Delta T$	$\Delta E_{th} = Q$
Isobaric	$V_i/T_i = V_f/T_f$	$p\,\Delta V$	$nC_P\Delta T$	$\Delta E_{th} = Q - W_s$
Isothermal	$p_iV_i = p_fV_f$	$nRT \ln(V_f/V_i)$ $pV \ln(V_f/V_i)$	$Q = W_s$	$\Delta E_{th} = 0$
Adiabatic	$p_iV_i^\gamma = p_fV_f^\gamma$ $T_iV_i^{\gamma-1} = T_fV_f^{\gamma-1}$	$(p_fV_f - p_iV_i)/(1-\gamma)$ $-nC_V\Delta T$	0	$\Delta E_{th} = -W_s$
Any	$p_iV_i/T_i = p_fV_f/T_f$	area under curve		$\Delta E_{th} = nC_V\Delta T$

TABLE 19.2 Properties of monatomic and diatomic gases

	Monatomic	Diatomic
E_{th}	$\frac{3}{2}nRT$	$\frac{5}{2}nRT$
C_V	$\frac{3}{2}R$	$\frac{5}{2}R$
C_P	$\frac{5}{2}R$	$\frac{7}{2}R$
γ	$\frac{5}{3} = 1.67$	$\frac{7}{5} = 1.40$

There is one entry in this table that you haven't seen before. The expression

$$W_s = \frac{p_fV_f - p_iV_i}{1-\gamma} \qquad \text{(work in an adiabatic process)} \qquad (19.12)$$

for the work done in an adiabatic process follows from writing $W_s = -\Delta E_{th} = -nC_V\Delta T$, which you learned in Chapter 17, then using $\Delta T = \Delta(pV)/nR$ and the definition of γ. The proof will be left for a homework problem.

You learned in Chapter 18 that the thermal energy of an ideal gas depends only on its temperature. Table 19.2 lists the thermal energy, molar specific heats, and specific heat ratio $\gamma = C_P/C_V$ for monatomic and diatomic gases.

A Strategy for Heat-Engine Problems

8.12, 8.13 Active Physics ONLINE

The engine of Example 19.1 was not a realistic heat engine, but it did illustrate the kinds of reasoning and computations involved in the analysis of a heat engine. A basic strategy for analyzing a heat engine follows.

PROBLEM-SOLVING STRATEGY 19.1 **Heat-engine problems**

MODEL Identify each process in the cycle.

VISUALIZE Draw the pV diagram of the cycle.

SOLVE There are several steps in the mathematical analysis.

- Use the ideal-gas law to complete your knowledge of n, p, V, and T at one point in the cycle.
- Use the ideal-gas law and equations for specific gas processes to determine p, V, and T at the beginning and end of each process.
- Calculate Q, W_s, and ΔE_{th} for each process.
- Find W_{out} by adding W_s for each process in the cycle. If the geometry is simple, you can confirm this value by finding the area enclosed within the pV curve.
- Add just the *positive* values of Q to find Q_H.
- Verify that $(\Delta E_{th})_{net} = 0$. This is a self-consistency check to verify that you haven't made any mistakes.
- Calculate the thermal efficiency η and any other quantities you need to complete the solution.

ASSESS Is $(\Delta E_{th})_{net} = 0$? Do all the signs of W_s and Q make sense? Does η have a reasonable value? Have you answered the question?

EXAMPLE 19.2 Analyzing a heat engine II

A heat engine with a diatomic gas as the working substance uses the closed cycle shown in **FIGURE 19.14**. How much work does this engine do per cycle, and what is its thermal efficiency?

FIGURE 19.14 The pV diagram for the heat engine of Example 19.2.

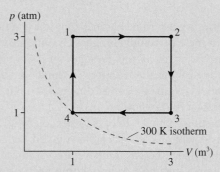

MODEL Processes $1 \rightarrow 2$ and $3 \rightarrow 4$ are isobaric. Processes $2 \rightarrow 3$ and $4 \rightarrow 1$ are isochoric.

VISUALIZE The pV diagram has already been drawn.

SOLVE We know the pressure, volume, and temperature at state 4. The number of moles of gas in the heat engine is

$$n = \frac{p_4 V_4}{RT_4} = \frac{(101{,}300 \text{ Pa})(1.0 \text{ m}^3)}{(8.31 \text{ J/mol K})(300 \text{ K})} = 40.6 \text{ mol}$$

p/T = constant during an isochoric process and V/T = constant during an isobaric process. These allow us to find that $T_1 = T_3 = 900 \text{ K}$ and $T_2 = 2700 \text{ K}$. This completes our knowledge of the state variables at all four corners of the diagram.

Process $1 \rightarrow 2$ is an isobaric expansion, so

$$(W_s)_{12} = p\Delta V = (3.0 \times 101{,}300 \text{ Pa})(2.0 \text{ m}^3) = 6.08 \times 10^5 \text{ J}$$

where we converted the pressure to pascals. The heat transfer during an isobaric expansion is

$$Q_{12} = nC_P\Delta T = (40.6 \text{ mol})(29.1 \text{ J/mol K})(1800 \text{ K})$$
$$= 21.27 \times 10^5 \text{ J}$$

where $C_P = \frac{7}{2}R$ for a diatomic gas. Then, using the first law,

$$\Delta E_{12} = Q_{12} - (W_s)_{12} = 15.19 \times 10^5 \text{ J}$$

Process $2 \rightarrow 3$ is an isochoric process, so $(W_s)_{23} = 0$ and

$$\Delta E_{23} = Q_{23} = nC_V\Delta T = -15.19 \times 10^5 \text{ J}$$

Notice that ΔT is *negative*.

Process $3 \rightarrow 4$ is an isobaric compression. Now ΔV is negative, so

$$(W_s)_{34} = p\Delta V = -2.03 \times 10^5 \text{ J}$$

and

$$Q_{34} = nC_P\Delta T = -7.09 \times 10^5 \text{ J}$$

Then $\Delta E_{th} = Q_{34} - (W_s)_{34} = -5.06 \times 10^5 \text{ J}$.

Process $4 \rightarrow 1$ is another constant-volume process, so again $(W_s)_{41} = 0$ and

$$\Delta E_{41} = Q_{41} = nC_V\Delta T = 5.06 \times 10^5 \text{ J}$$

The results of all four processes are shown in Table 19.3. The net results for W_{out}, Q_{net}, and $(\Delta E_{th})_{net}$ are found by summing the columns. As expected, $W_{out} = Q_{net}$ and $(\Delta E_{th})_{net} = 0$.

TABLE 19.3 Energy transfers in Example 19.2. All energies $\times 10^5$ J

Process	W_s	Q	ΔE_{th}
$1 \rightarrow 2$	6.08	21.27	15.19
$2 \rightarrow 3$	0	−15.19	−15.19
$3 \rightarrow 4$	−2.03	−7.09	−5.06
$4 \rightarrow 1$	0	5.06	5.06
Net	4.05	4.05	0

The work done during one cycle is $W_{out} = 4.05 \times 10^5$ J. Heat enters the system from the hot reservoir during processes $1 \rightarrow 2$ and $4 \rightarrow 1$, where Q is positive. Summing these gives $Q_H = 26.33 \times 10^5$ J. Thus the thermal efficiency of this engine is

$$\eta = \frac{W_{out}}{Q_H} = \frac{4.05 \times 10^5 \text{ J}}{26.33 \times 10^5 \text{ J}} = 0.15 = 15\%$$

ASSESS The verification that $W_{out} = Q_{net}$ and $(\Delta E_{th})_{net} = 0$ gives us great confidence that we didn't make any calculational errors. This engine may not seem very efficient, but η is quite typical of many real engines.

We noted in Example 19.1 that a heat engine's hot-reservoir temperature T_H must exceed the highest temperature reached by the system and the cold-reservoir temperature T_C must be less than the coldest system temperature. Although we don't know what the reservoirs are in Example 19.2, we can be sure that $T_H > 2700$ K and $T_C < 300$ K.

STOP TO THINK 19.3 What is the thermal efficiency of this heat engine?

a. 0.10
b. 0.50
c. 0.25
d. 4
e. Can't tell without knowing Q_C.

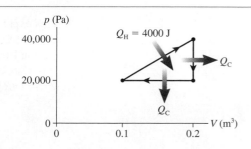

A jet engine uses a modified Brayton cycle.

FIGURE 19.15 A gas turbine engine follows a Brayton cycle.

(a)

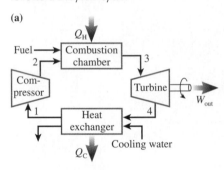

(b)

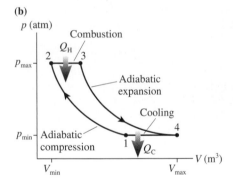

The Brayton Cycle

The heat engines of Examples 19.1 and 19.2 have been educational but not realistic. As an example of a more realistic heat engine we'll look at the thermodynamic cycle known as the *Brayton cycle*. It is a reasonable model of a *gas turbine engine*. Gas turbines are used for electric power generation and as the basis for jet engines in aircraft and rockets. The *Otto cycle*, which describes the gasoline internal combustion engine, and the *Diesel cycle*, which, not surprisingly, describes the diesel engine, will be the subject of homework problems.

FIGURE 19.15a is a schematic look at a gas turbine engine, and **FIGURE 19.15b** is the corresponding pV diagram. To begin the Brayton cycle, air at an initial pressure p_1 is rapidly compressed in a *compressor*. This is an *adiabatic process*, with $Q = 0$, because there is no time for heat to be exchanged with the surroundings. Recall that an adiabatic compression raises the temperature of a gas by doing work on it, not by heating it, so the air leaving the compressor is very hot.

The hot gas flows into a combustion chamber. Fuel is continuously admitted to the combustion chamber where it mixes with the hot gas and is ignited, transferring heat to the gas at constant pressure and raising the gas temperature yet further. The high-pressure gas then expands, spinning a turbine that does some form of useful work. This adiabatic expansion, with $Q = 0$, drops the temperature and pressure of the gas. The pressure at the end of the expansion through the turbine is back to p_1, but the gas is still quite hot. The gas completes the cycle by flowing through a device called a **heat exchanger** that transfers heat energy to a cooling fluid. Large power plants are often sited on rivers or oceans in order to use the water for the cooling fluid in the heat exchanger.

This thermodynamic cycle, called a Brayton cycle, has two adiabatic processes—the compression and the expansion through the turbine—plus a constant-pressure heating and a constant-pressure cooling. There's no heat transfer during the adiabatic processes. The hot-reservoir temperature must be $T_H \geq T_3$ for heat to be transferred into the gas during process $2 \to 3$. Similarly, the heat exchanger will remove heat from the gas only if $T_C \leq T_1$.

The thermal efficiency of any heat engine is

$$\eta = \frac{W_{out}}{Q_H} = 1 - \frac{Q_C}{Q_H}$$

Heat is transferred into the gas only during process $2 \to 3$. This is an isobaric process, so $Q_H = nC_P\Delta T = nC_P(T_3 - T_2)$. Similarly, heat is transferred out only during the isobaric process $4 \to 1$.

We have to be careful with signs. Q_{41} is negative because the temperature decreases, but Q_C was defined as the *amount* of heat exchanged with the cold reservoir, a positive quantity. Thus

$$Q_C = |Q_{41}| = |nC_P(T_1 - T_4)| = nC_P(T_4 - T_1) \qquad (19.13)$$

With these expressions for Q_H and Q_C, the thermal efficiency is

$$\eta_{Brayton} = 1 - \frac{T_4 - T_1}{T_3 - T_2} \qquad (19.14)$$

This expression isn't useful unless we compute all four temperatures. Fortunately, we can cast Equation 19.14 into a more useful form.

You learned in Chapter 17 that pV^γ = constant during an adiabatic process, where $\gamma = C_P/C_V$ is the specific heat ratio. If we use $V = nRT/p$ from the ideal-gas law, $V^\gamma = (nR)^\gamma T^\gamma p^{-\gamma}$. $(nR)^\gamma$ is a constant, so we can write pV^γ = constant as

$$p^{(1-\gamma)}T^\gamma = \text{constant} \qquad (19.15)$$

Equation 19.15 is a pressure-temperature relationship for an adiabatic process. Because $(T^\gamma)^{1/\gamma} = T$, we can simplify Equation 19.15 by raising both sides to the power $1/\gamma$. Doing so gives

$$p^{(1-\gamma)/\gamma}T = \text{constant} \tag{19.16}$$

during an adiabatic process.

Process $1 \rightarrow 2$ is an adiabatic process; hence

$$p_1^{(1-\gamma)/\gamma}T_1 = p_2^{(1-\gamma)/\gamma}T_2 \tag{19.17}$$

Isolating T_1 gives

$$T_1 = \frac{p_2^{(1-\gamma)/\gamma}}{p_1^{(1-\gamma)/\gamma}}T_2 = \left(\frac{p_2}{p_1}\right)^{(1-\gamma)/\gamma}T_2 = \left(\frac{p_{max}}{p_{min}}\right)^{(1-\gamma)/\gamma}T_2 \tag{19.18}$$

If we define the **pressure ratio** r_p as $r_p = p_{max}/p_{min}$, then T_1 and T_2 are related by

$$T_1 = r_p^{(1-\gamma)/\gamma}T_2 \tag{19.19}$$

The algebra of getting to Equation 19.19 was a bit tricky, but the final result is fairly simple.

Process $3 \rightarrow 4$ is also an adiabatic process. The same reasoning leads to

$$T_4 = r_p^{(1-\gamma)/\gamma}T_3 \tag{19.20}$$

If we substitute these expressions for T_1 and T_4 into Equation 19.14, the efficiency is

$$\eta_B = 1 - \frac{T_4 - T_1}{T_3 - T_2} = 1 - \frac{r_p^{(1-\gamma)/\gamma}T_3 - r_p^{(1-\gamma)/\gamma}T_2}{T_3 - T_2} = 1 - \frac{r_p^{(1-\gamma)/\gamma}(T_3 - T_2)}{T_3 - T_2}$$

$$= 1 - r_p^{(1-\gamma)/\gamma}$$

Remarkably, all the temperatures cancel and we're left with an expression that depends only on the pressure ratio. Noting that $(1 - \gamma)$ is negative, we can make one final change and write

$$\eta_B = 1 - \frac{1}{r_p^{(\gamma-1)/\gamma}} \tag{19.21}$$

FIGURE 19.16 is a graph of the efficiency of the Brayton cycle as a function of the pressure ratio, assuming $\gamma = 1.40$ for a diatomic gas such as air.

In Example 19.2 we found the thermal efficiency $\eta = W_{out}/Q_H$ by explicitly computing W_{out} and Q_H. Here, by contrast, we've determined the thermal efficiency of the Brayton cycle by using the relationship between the initial and final temperatures during an adiabatic process. The price we pay for this simplified analysis is that we didn't find an expression for the work done by a heat engine following the Brayton cycle. To calculate the work, which you can do as a homework problem, there's no avoiding the step-by-step analysis of the problem-solving strategy.

FIGURE 19.16 The efficiency of a Brayton cycle as a function of the pressure ratio r_p.

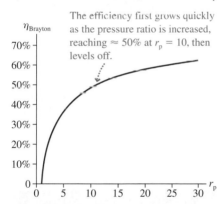

The efficiency first grows quickly as the pressure ratio is increased, reaching $\approx 50\%$ at $r_p = 10$, then levels off.

Any increase in efficiency beyond $\approx 50\%$ has to be weighed against the higher costs of a better compressor that can achieve a much higher pressure ratio.

19.4 Ideal-Gas Refrigerators

Suppose we were to operate a Brayton heat engine backward, going *ccw* in the *pV* diagram. **FIGURE 19.17a**, on the next page, (which you should compare to Figure 19.15a) shows a device for doing this. **FIGURE 19.17b** is its *pV* diagram, and **FIGURE 19.17c** is the energy-transfer diagram. Starting from point 4, the gas is adiabatically compressed to increase its temperature and pressure. It then flows through a high-temperature heat exchanger where the gas *cools* at constant pressure from temperature T_3 to T_2. The gas then expands adiabatically, leaving it significantly colder at T_1 than it started at T_4. It

FIGURE 19.17 A refrigerator that extracts heat from the cold reservoir and exhausts heat to the hot reservoir.

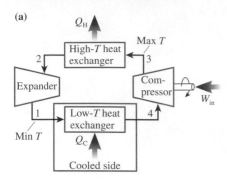

(a)

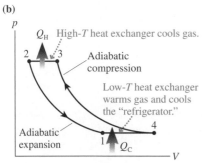

(b)

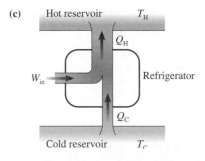

(c)

These cooling coils are the refrigerator's high-temperature heat exchanger. Heat energy is being transferred from hot gas inside the coils to the cooler room air.

completes the cycle by flowing through a low-temperature heat exchanger, where it *warms* back to its starting temperature.

Suppose that the low-temperature heat exchanger is a closed container of air surrounding a pipe through which the engine's cold gas is flowing. The heat-exchange process $1 \rightarrow 4$ *cools* the air in the container as it warms the gas flowing through the pipe. If you were to place eggs and milk inside this closed container, you would call it a refrigerator!

Going around a closed pV cycle in a ccw direction reverses the sign of W for each process in the cycle. Consequently, the area inside the curve of Figure 19.17b is W_{in}, the work done *on* the system. Here work is used to extract heat Q_C from the cold reservoir and exhaust a larger amount of heat $Q_H = Q_C + W_{in}$ to the hot reservoir. But where, in this situation, are the energy reservoirs?

Understanding a refrigerator is a little harder than understanding a heat engine. The key is to remember that **heat is always transferred from a hotter object to a colder object.** In particular,

- The gas in a refrigerator can extract heat Q_C *from* the cold reservoir only if the gas temperature is *lower* than the cold-reservoir temperature T_C. Heat energy is then transferred *from* the cold reservoir *into* the colder gas.
- The gas in a refrigerator can exhaust heat Q_H *to* the hot reservoir only if the gas temperature is *higher* than the hot-reservoir temperature T_H. Heat energy is then transferred *from* the warmer gas *into* the hot reservoir.

These two requirements place severe constraints on the thermodynamics of a refrigerator. Because there is no reservoir colder than T_C, the gas cannot reach a temperature lower than T_C by heat exchange. The gas in a refrigerator *must* use an adiabatic expansion ($Q = 0$) to lower the temperature below T_C. Likewise, a gas refrigerator requires an adiabatic compression to raise the gas temperature above T_H.

The reversed Brayton cycle of Figure 19.17b does, indeed, have two adiabatic processes. The adiabatic expansion lowers the temperature to T_1, then heat Q_C is transferred *from* the cold reservoir *to* the gas during process $1 \rightarrow 4$. Consequently, the cold-reservoir temperature must be $T_C \geq T_4$. Contrast this with the same cycle run clockwise (cw) as a heat engine, where we saw that the cold reservoir must be $T_C \leq T_1$.

Similar reasoning applies on the hot side. In order for heat Q_H to be transferred *into* the hot reservoir during process $3 \rightarrow 2$, the hot-reservoir temperature must be $T_H \leq T_2$. This requirement of the high-temperature reservoir differs distinctly from the Brayton heat engine, which required $T_H \geq T_3$. **FIGURE 19.18** compares a Brayton-cycle heat engine to a Brayton-cycle refrigerator.

FIGURE 19.18 A comparison of a Brayton-cycle heat engine to a Brayton-cycle refrigerator.

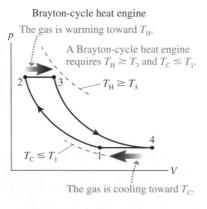

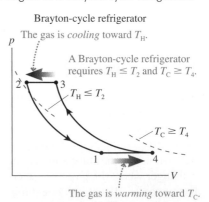

The important point—a point we will return to in the next section—is that a Brayton refrigerator is *not* simply a Brayton heat engine running backward. To make a Brayton refrigerator you must both reverse the cycle *and* change the hot and cold reservoirs.

NOTE ▶ Some heat engines cannot be converted to refrigerators under any circumstances. We'll leave it as a homework problem to show that the heat engine of Example 19.2, if run backward, is a total loser. Its energy-transfer diagram, shown in FIGURE 19.19, shows work being done to transfer energy "downhill" even faster than it would move spontaneously from hot to cold! ◀

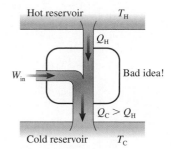

FIGURE 19.19 This is the energy-transfer diagram if the heat engine of Example 19.2 is run backward.

EXAMPLE 19.3 Analyzing a refrigerator

A refrigerator using helium gas operates on a reversed Brayton cycle with a pressure ratio of 5.0. Prior to compression, the gas occupies 100 cm³ at a pressure of 150 kPa and a temperature of $-23°C$. Its volume at the end of the expansion is 80 cm³. What are the refrigerator's coefficient of performance and its power input if it operates at 60 cycles per second?

MODEL The Brayton cycle has two adiabatic processes and two isobaric processes. The work per cycle needed to run the refrigerator is $W_{in} = Q_H - Q_C$; hence we can determine both the coefficient of performance and the power requirements from Q_H and Q_C. Heat energy is transferred only during the two isobaric processes.

VISUALIZE FIGURE 19.20 shows the pV cycle. We know from the pressure ratio of 5.0 that the maximum pressure is 750 kPa. Neither V_2 nor V_3 is known.

FIGURE 19.20 A Brayton-cycle refrigerator.

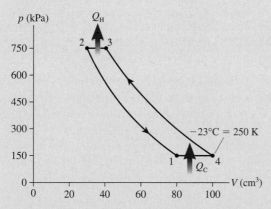

SOLVE To calculate heat we're going to need the temperatures at the four corners of the cycle. First, we can use the conditions of state 4 to find the number of moles of helium:

$$n = \frac{p_4 V_4}{R T_4} = 0.00722 \text{ mol}$$

Process $1 \rightarrow 4$ is isobaric; hence temperature T_1 is

$$T_1 = \frac{V_1}{V_4} T_4 = (0.80)(250 \text{ K}) = 200 \text{ K} = -73°C$$

With Equation 19.16 we found that the quantity $p^{(1-\gamma)/\gamma}T$ remains constant during an adiabatic process. Helium is a monatomic gas with $\gamma = \frac{5}{3}$, so $(1 - \gamma)/\gamma = -\frac{2}{5} = -0.40$. For the adiabatic compression $4 \rightarrow 3$,

$$p_3^{-0.40} T_3 = p_4^{-0.40} T_4$$

Solving for T_3 gives

$$T_3 = \left(\frac{p_4}{p_3}\right)^{-0.40} T_4 = \left(\frac{1}{5}\right)^{-0.40} (250 \text{ K}) = 476 \text{ K} = 203°C$$

The same analysis applied to the $2 \rightarrow 1$ adiabatic expansion gives

$$T_2 = \left(\frac{p_1}{p_2}\right)^{-0.40} T_1 = \left(\frac{1}{5}\right)^{-0.40} (200 \text{ K}) = 381 \text{ K} = 108°C$$

Now we can use $C_P = \frac{5}{2}R = 20.8 \text{ J/mol K}$ for a monatomic gas to compute the heat transfers:

$$Q_H = |Q_{32}| = nC_P(T_3 - T_2)$$

$$= (0.00722 \text{ mol})(20.8 \text{ J/mol K})(95 \text{ K}) = 14.3 \text{ J}$$

$$Q_C = |Q_{14}| = nC_P(T_4 - T_1)$$

$$= (0.00722 \text{ mol})(20.8 \text{ J/mol K})(50 \text{ K}) = 7.5 \text{ J}$$

Thus the work *input* to the refrigerator is $W_{in} = Q_H - Q_C = 6.8 \text{ J}$. During each cycle, 6.8 J of work are done *on* the gas to extract 7.5 J of heat from the cold reservoir. Then 14.3 J of heat are exhausted into the hot reservoir.

The refrigerator's coefficient of performance is

$$K = \frac{Q_C}{W_{in}} = \frac{7.5 \text{ J}}{6.8 \text{ J}} = 1.1$$

The power input needed to run the refrigerator is

$$P_{in} = 6.8 \frac{\text{J}}{\text{cycle}} \times 60 \frac{\text{cycles}}{\text{s}} = 410 \frac{\text{J}}{\text{s}} = 410 \text{ W}$$

ASSESS These are fairly realistic values for a kitchen refrigerator. You pay your electric company for providing the work W_{in} that operates the refrigerator. The cold reservoir is the freezer compartment. The cold temperature T_C must be higher than T_4 ($T_C > -23°C$) in order for heat to be transferred *from* the cold reservoir *to* the gas. A typical freezer temperature is $-15°C$, so this condition is satisfied. The hot reservoir is the air in the room. The back and underside of a refrigerator have heat-exchanger coils where the hot gas, after compression, transfers heat to the air. The hot temperature T_H must be less than T_2 ($T_H < 108°C$) in order for heat to be transferred *from* the gas *to* the air. An air temperature $\approx 25°C$ under a refrigerator satisfies this condition.

What, if anything, is wrong with this refrigerator?

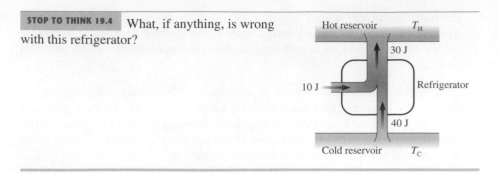

19.5 The Limits of Efficiency

Everyone knows that heat can produce motion. That it possesses vast motive power no one can doubt, in these days when the steam engine is everywhere so well known. . . . Notwithstanding the satisfactory condition to which they have been brought today, their theory is very little understood. The question has often been raised whether the motive power of heat is unbounded, or whether the possible improvements in steam engines have an assignable limit.

Sadi Carnot

Thermodynamics has its historical roots in the development of the steam engine and other machines of the early industrial revolution. Early steam engines, built on the basis of experience rather than scientific understanding, were not very efficient at converting fuel energy into work. The first major theoretical analysis of heat engines was published by the French engineer Sadi Carnot in 1824. The question that Carnot raised was one we posed at the end of Section 19.3: Can we make a heat engine whose thermal efficiency η approaches 1, or is there an upper limit η_{max} that cannot be exceeded? To frame the question more clearly, imagine we have a hot reservoir at temperature T_H and a cold reservoir at T_C. What is the most efficient heat engine (maximum η) that can operate between these two energy reservoirs? Similarly, what is the most efficient refrigerator (maximum K) that can operate between the two reservoirs?

We just saw that a refrigerator is, in some sense, a heat engine running backward. We might thus suspect that the most efficient heat engine is related to the most efficient refrigerator. Suppose we have a heat engine that we can turn into a refrigerator by reversing the direction of operation, thus changing the direction of the energy transfers, and with *no other changes*. In particular, the heat engine and the refrigerator operate between the same two energy reservoirs at temperatures T_H and T_C.

NOTE ▶ The heat engine we're looking for cannot be a Brayton-cycle heat engine. A Brayton-cycle refrigerator requires reversing the heat engine's direction of operation *and* changing the temperatures of the energy reservoirs. ◀

FIGURE 19.21a shows such a heat engine and its corresponding refrigerator. Notice that the refrigerator has *exactly the same* work and heat transfer as the heat engine, only in the opposite directions. A device that can be operated as either a heat engine or a refrigerator between the same two energy reservoirs and with the same energy transfers, with only their direction changed, is called a **perfectly reversible engine.** A perfectly reversible engine is an idealization, as was the concept of a perfectly elastic collision. Nonetheless, it will allow us to establish limits that no real engine can exceed.

FIGURE 19.21 If a perfectly reversible heat engine is used to operate a perfectly reversible refrigerator, the two devices exactly cancel each other.

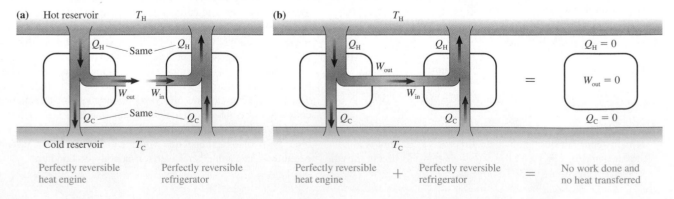

Suppose we have a perfectly reversible heat engine and a perfectly reversible refrigerator (the same device running backward) operating between a hot reservoir at temperature T_H and a cold reservoir at temperature T_C. Because the work W_{in} needed to operate the refrigerator is exactly the same as the useful work W_{out} done by the heat engine, we can use the heat engine, as shown in **FIGURE 19.21b**, to drive the refrigerator. The heat Q_C the engine exhausts to the cold reservoir is exactly the same as the heat Q_C the refrigerator extracts from the cold reservoir. Similarly, the heat Q_H the engine extracts from the hot reservoir matches the heat Q_H the refrigerator exhausts to the hot reservoir. Consequently, there is no net heat transfer in either direction. The refrigerator exactly replaces all the heat energy that had been transferred out of the hot reservoir by the heat engine.

You may want to compare the reasoning used here with the reasoning we used with Figure 19.12. There we tried to use the output of a "perfect" heat engine to run a refrigerator but did *not* succeed.

A Perfectly Reversible Engine Has Maximum Efficiency

Now we've arrived at the critical step in the reasoning. Suppose I claim to have a heat engine that can operate between temperatures T_H and T_C with *more* efficiency than a perfectly reversible engine. **FIGURE 19.22** shows the output of this heat engine operating the same perfectly reversible refrigerator that we used in Figure 19.21b.

FIGURE 19.22 A heat engine more efficient than a perfectly reversible engine could be used to violate the second law of thermodynamics.

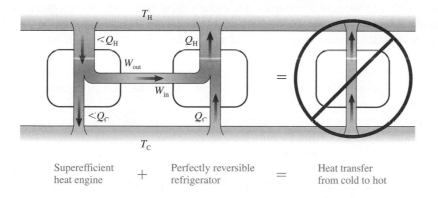

| Superefficient heat engine | $+$ | Perfectly reversible refrigerator | $=$ | Heat transfer from cold to hot |

Recall that the thermal efficiency and the work of a heat engine are

$$\eta = \frac{W_{out}}{Q_H} \quad \text{and} \quad W_{out} = Q_H - Q_C$$

If the new heat engine is more efficient than the perfectly reversible engine it replaces, it needs *less* heat Q_H from the hot reservoir to perform the *same* work W_{out}. If Q_H is less while W_{out} is the same, then Q_C must also be less. That is, the new heat engine exhausts less heat to the cold reservoir than does the perfectly reversible heat engine.

When this new heat engine drives the perfectly reversible refrigerator, the heat it exhausts to the cold reservoir is *less* than the heat extracted from the cold reservoir by the refrigerator. Similarly, this engine extracts *less* heat from the hot reservoir than the refrigerator exhausts. Thus the net result of using this superefficient heat engine to operate a perfectly reversible refrigerator is that heat is transferred from the cold reservoir to the hot reservoir *without outside assistance*.

But this can't happen. It would violate the second law of thermodynamics. Hence we have to conclude that no heat engine operating between reservoirs at temperatures T_H and T_C can be more efficient than a perfectly reversible engine. This very important conclusion is another version of the second law:

> **Second law, informal statement #5** No heat engine operating between reservoirs at temperatures T_H and T_C can be more efficient than a perfectly reversible engine operating between these temperatures.

The answer to our question "Is there a maximum η that cannot be exceeded?" is a clear "Yes!" The maximum possible efficiency η_{max} is that of a perfectly reversible engine. Because the perfectly reversible engine is an idealization, any real engine will have an efficiency less than η_{max}.

A similar argument shows that no refrigerator can be more efficient than a perfectly reversible refrigerator. If we had such a refrigerator, and if we ran it with the output of a perfectly reversible heat engine, we could transfer heat from cold to hot with no outside assistance. Thus:

> **Second law, informal statement #6** No refrigerator operating between reservoirs at temperatures T_H and T_C can have a coefficient of performance larger than that of a perfectly reversible refrigerator operating between these temperatures.

Conditions for a Perfectly Reversible Engine

This argument tells us that η_{max} and K_{max} exist, but it doesn't tell us what they are. Our final task will be to "design" and analyze a perfectly reversible engine. Under what conditions is an engine reversible?

An engine transfers energy by both mechanical and thermal interactions. Mechanical interactions are pushes and pulls. The environment does work on the system, transferring energy into the system by pushing in on a piston. The system transfers energy back to the environment by pushing out on the piston.

The energy transferred by a moving piston is perfectly reversible, returning the system to its initial state, with no change of temperature or pressure, only if the motion is *frictionless.* The slightest bit of friction will prevent the mechanical transfer of energy from being perfectly reversible.

The circumstances under which heat transfer can be *completely* reversed aren't quite so obvious. After all, Chapter 18 emphasized the *irreversible* nature of heat transfer. If objects A and B are in thermal contact, with $T_A > T_B$, then heat energy is transferred from A to B. But the second law of thermodynamics prohibits a heat transfer from B back to A. Heat transfer through a temperature *difference* is an irreversible process.

But suppose $T_A = T_B$. With no temperature difference, any heat that is transferred from A to B can, at a later time, be transferred from B back to A. This transfer wouldn't violate the second law, which prohibits only heat transfer from a colder object to a hotter object. Now you might object, and rightly so, that heat *can't* move from A to B if they are at the same temperature because heat, by definition, is the energy transferred between two objects at different temperatures.

This is true, so let's consider a limiting case in which $T_A = T_B + dT$. The temperature difference is infinitesimal. Heat is transferred from A to B, but *very slowly!* If you later try to make the heat move from B back to A, the second law will prevent you from doing so with perfect precision. But because the temperature difference is infinitesimal, you'll be missing only an infinitesimal amount dQ of heat. You can transfer heat reversibly in the limit $dT \rightarrow 0$, but you must be prepared to spend an infinite amount of time doing so.

Thus the thermal transfer of energy is reversible if the heat is transferred infinitely slowly in an isothermal process. This is an idealization, but so are completely frictionless processes. Nonetheless, we can now say that a perfectly reversible engine must use only two types of processes:

1. Frictionless mechanical interactions with no heat transfer ($Q = 0$), and
2. Thermal interactions in which heat is transferred in an isothermal process ($\Delta E_{th} = 0$).

Any engine that uses only these two types of processes is called a **Carnot engine.** A Carnot engine is a perfectly reversible engine; thus it has the maximum possible thermal efficiency η_{max} and, if operated as a refrigerator, the maximum possible coefficient of performance K_{max}.

19.6 The Carnot Cycle

No real engine is perfectly reversible, so a Carnot engine is an idealization. Nonetheless, an analysis of the Carnot engine will allow us to establish a maximum possible thermal efficiency that no real heat engine can exceed.

 8.14

The definition of a Carnot engine does not specify whether the engine's working substance is a gas or a liquid. It makes no difference. Our argument that a perfectly reversible engine is the most efficient possible heat engine depended only on the engine's reversibility. It did not depend on any details of how the engine is constructed or what it uses for a working substance. Consequently, **any Carnot engine operating between T_H and T_C must have exactly the same efficiency as any other Carnot engine operating between the same two energy reservoirs.** If we can determine the thermal efficiency of one Carnot engine, we'll know the efficiency of all Carnot engines. Because liquids and phase changes are complicated, we'll analyze a Carnot engine that uses an ideal gas.

The Carnot Cycle

The **Carnot cycle** is an ideal-gas cycle that consists of the two adiabatic processes ($Q = 0$) and two isothermal processes ($\Delta E_{th} = 0$) shown in **FIGURE 19.23**. These are the two types of processes allowed in a perfectly reversible gas engine. As a Carnot cycle operates,

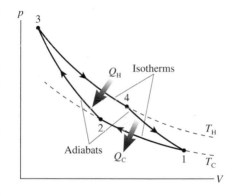

FIGURE 19.23 The Carnot cycle is perfectly reversible.

1. The gas is isothermally compressed while in thermal contact with the cold reservoir at temperature T_C. Heat energy $Q_C = |Q_{12}|$ is removed from the gas as it is compressed in order to keep the temperature constant. The compression must take place extremely slowly because there can be only an infinitesimal temperature difference between the gas and the reservoir.
2. The gas is adiabatically compressed while thermally isolated from the environment. This compression increases the gas temperature until it matches temperature T_H of the hot reservoir. No heat is transferred during this process.
3. After reaching maximum compression, the gas expands isothermally at temperature T_H. Heat $Q_H = Q_{34}$ is transferred from the hot reservoir into the gas as it expands in order to keep the temperature constant.
4. Finally, the gas expands adiabatically, with $Q = 0$, until the temperature decreases back to T_C

Work is done in all four processes of the Carnot cycle, but heat is transferred only during the two isothermal processes.

The thermal efficiency of any heat engine is

$$\eta = \frac{W_{out}}{Q_H} = 1 - \frac{Q_C}{Q_H}$$

We can determine η_{Carnot} by finding the heat transfer in the two isothermal processes.

Process $1 \rightarrow 2$: Table 19.1 gives us the heat transfer in an isothermal process at temperature T_C:

$$Q_{12} = (W_s)_{12} = nRT_C \ln\left(\frac{V_2}{V_1}\right) = -nRT_C \ln\left(\frac{V_1}{V_2}\right) \qquad (19.22)$$

$V_1 > V_2$, so the logarithm on the right is positive. Q_{12} is negative because heat is transferred out of the system, but Q_C is simply the *amount* of heat transferred to the cold reservoir:

$$Q_C = |Q_{12}| = nRT_C \ln\left(\frac{V_1}{V_2}\right) \tag{19.23}$$

Process $3 \rightarrow 4$: Similarly, the heat transferred in the isothermal expansion at temperature T_H is

$$Q_H = Q_{34} = (W_s)_{34} = nRT_H \ln\frac{V_4}{V_3} \tag{19.24}$$

Thus the thermal efficiency of the Carnot cycle is

$$\eta_{Carnot} = 1 - \frac{Q_C}{Q_H} = 1 - \frac{T_C \ln(V_1/V_2)}{T_H \ln(V_4/V_3)} \tag{19.25}$$

We can simplify this expression. During the two adiabatic processes,

$$T_C V_2^{\gamma-1} = T_H V_3^{\gamma-1} \quad \text{and} \quad T_C V_1^{\gamma-1} = T_H V_4^{\gamma-1} \tag{19.26}$$

An algebraic rearrangement gives

$$V_2 = V_3\left(\frac{T_H}{T_C}\right)^{1/(\gamma-1)} \quad \text{and} \quad V_1 = V_4\left(\frac{T_H}{T_C}\right)^{1/(\gamma-1)} \tag{19.27}$$

from which it follows that

$$\frac{V_1}{V_2} = \frac{V_4}{V_3} \tag{19.28}$$

Consequently, the two logarithms in Equation 19.25 cancel and we're left with the result that the thermal efficiency of a Carnot engine operating between a hot reservoir at temperature T_H and a cold reservoir at temperature T_C is

$$\eta_{Carnot} = 1 - \frac{T_C}{T_H} \quad \text{(Carnot thermal efficiency)} \tag{19.29}$$

This remarkably simple result, an efficiency that depends only on the ratio of the temperatures of the hot and cold reservoirs, is Carnot's legacy to thermodynamics.

NOTE ▶ Temperatures T_H and T_C are *absolute* temperatures. ◀

EXAMPLE 19.4 **A Carnot engine**
A Carnot engine is cooled by water at $T_C = 10°C$. What temperature must be maintained in the hot reservoir of the engine to have a thermal efficiency of 70%?

MODEL The efficiency of a Carnot engine depends only on the temperatures of the hot and cold reservoirs.

SOLVE The thermal efficiency $\eta_{Carnot} = 1 - T_C/T_H$ can be rearranged to give

$$T_H = \frac{T_C}{1 - \eta_{Carnot}} = 943 \text{ K} = 670°C$$

ASSESS A "real" engine would need a higher temperature than this to provide 70% efficiency because no real engine will match the Carnot efficiency.

EXAMPLE 19.5 **A real engine**
The heat engine of Example 19.2 had a highest temperature of 2700 K, a lowest temperature of 300 K, and a thermal efficiency of 15%. What is the efficiency of a Carnot engine operating between these two temperatures?

SOLVE The Carnot efficiency is

$$\eta_{Carnot} = 1 - \frac{T_C}{T_H} = 1 - \frac{300 \text{ K}}{2700 \text{ K}} = 0.89 = 89\%$$

ASSESS The thermodynamic cycle used in the example doesn't come anywhere close to the Carnot efficiency.

The Maximum Efficiency

In Section 19.2 we tried to invent a perfect engine with $\eta = 1$ and $Q_C = 0$. We found that we could not do so without violating the second law, so no engine can have $\eta = 1$. However, that example didn't rule out an engine with $\eta = 0.9999$. Further analysis has now shown that no heat engine operating between energy reservoirs at temperatures T_H and T_C can be more efficient than a perfectly reversible engine operating between these temperatures.

We've now reached the endpoint of this line of reasoning by establishing an exact result for the thermal efficiency of a perfectly reversible engine, the Carnot engine. We can summarize our conclusions:

> **Second law, informal statement #7** No heat engine operating between energy reservoirs at temperatures T_H and T_C can exceed the Carnot efficiency
> $$\eta_{\text{Carnot}} = 1 - \frac{T_C}{T_H}$$

As Example 19.5 showed, real engines usually fall well short of the Carnot limit.

We also found that no refrigerator can exceed the coefficient of performance of a perfectly reversible refrigerator. We'll leave the proof as a homework problem, but an analysis very similar to that above shows that the coefficient of performance of a Carnot refrigerator is

$$K_{\text{Carnot}} = \frac{T_C}{T_H - T_C} \qquad \text{(Carnot coefficient of performance)} \qquad (19.30)$$

Thus we can state:

> **Second law, informal statement #8** No refrigerator operating between energy reservoirs at temperatures T_H and T_C can exceed the Carnot coefficient of performance
> $$K_{\text{Carnot}} = \frac{T_C}{T_H - T_C}$$

EXAMPLE 19.6 Brayton versus Carnot

The Brayton-cycle refrigerator of Example 19.3 had coefficient of performance $K = 1.1$. Compare this to the limit set by the second law of thermodynamics.

SOLVE Example 19.3 found that the reservoir temperatures had to be $T_C \geq 250$ K and $T_H \leq 381$ K. A Carnot refrigerator operating between 250 K and 381 K has

$$K_{\text{Carnot}} = \frac{T_C}{T_H - T_C} = \frac{250 \text{ K}}{381 \text{ K} - 250 \text{ K}} = 1.9$$

ASSESS This is the minimum value of K_{Carnot}. It will be even higher if $T_C > 250$ K or $T_H < 381$ K. The coefficient of performance of the reasonably realistic refrigerator of Example 19.3 is less than 60% of the limiting value.

Statements #7 and #8 of the second law are a major result of this chapter, one with profound implications. The efficiency limit of a heat engine is set by the temperatures of the hot and cold reservoirs. High efficiency requires $T_C/T_H \ll 1$ and thus $T_H \gg T_C$. However, practical realities often prevent T_H from being significantly larger than T_C, in which case the engine cannot possibly have a large efficiency. This limit on the efficiency of heat engines is a consequence of the second law of thermodynamics.

EXAMPLE 19.7 Generating electricity

An electric power plant boils water to produce high-pressure steam at 400°C. The high-pressure steam spins a turbine as it expands, then the turbine spins the generator. The steam is then condensed back to water in an ocean-cooled heat exchanger at 25°C. What is the *maximum* possible efficiency with which heat energy can be converted to electric energy?

MODEL The maximum possible efficiency is that of a Carnot engine operating between these temperatures.

SOLVE The Carnot efficiency depends on absolute temperatures, so we must use $T_H = 400°C = 673$ K and $T_C = 25°C = 298$ K. Then

$$\eta_{max} = 1 - \frac{298}{673} = 0.56 = 56\%$$

ASSESS This is an upper limit. Real coal-, oil-, gas-, and nuclear-heated steam generators actually operate at $\approx 35\%$ thermal efficiency. (The heat *source* has nothing to do with the efficiency. All it does is boil water.) Thus, as in the photo at the beginning of this chapter, electric power plants convert only about one-third of the fuel energy to electric energy while exhausting about two-thirds of the energy to the environment as waste heat. Not much can be done to alter the low-temperature limit. The high-temperature limit is determined by the maximum temperature and pressure the boiler and turbine can withstand. The efficiency of electricity generation is far less than most people imagine, but it is an unavoidable consequence of the second law of thermodynamics.

A limit on the efficiency of heat engines was not expected. We are used to thinking in terms of energy conservation, so it comes as no surprise that we cannot make an engine with $\eta > 1$. But the limits arising from the second law were not anticipated, nor are they obvious. Nonetheless, they are a very real fact of life and a very real constraint on any practical device. No one has ever invented a machine that exceeds the second-law limits, and we have seen that the maximum efficiency for realistic engines is surprisingly low.

STOP TO THINK 19.5 Could this heat engine be built?

a. Yes.
b. No.
c. It's impossible to tell without knowing what kind of cycle it uses.

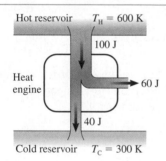

SUMMARY

The goal of Chapter 19 has been to study the physical principles that govern the operation of heat engines and refrigerators.

General Principles

Heat Engines

Devices that transform heat into work. They require two energy reservoirs at different temperatures.

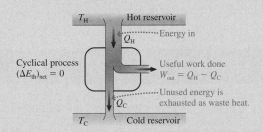

Thermal efficiency

$$\eta = \frac{W_{out}}{Q_H} = \frac{\text{what you get}}{\text{what you pay}}$$

Second-law limit:

$$\eta \leq 1 - \frac{T_C}{T_H}$$

Refrigerators

Devices that use work to transfer heat from a colder object to a hotter object.

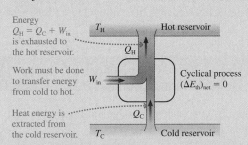

Energy $Q_H = Q_C + W_{in}$ is exhausted to the hot reservoir.

Work must be done to transfer energy from cold to hot.

Heat energy is extracted from the cold reservoir.

Coefficient of performance

$$K = \frac{Q_C}{W_{in}} = \frac{\text{what you get}}{\text{what you pay}}$$

Second-law limit:

$$K \leq \frac{T_C}{T_H - T_C}$$

Important Concepts

A perfectly reversible engine (a **Carnot engine**) can be operated as either a heat engine or a refrigerator between the same two energy reservoirs by reversing the cycle and with no other changes.

- A **Carnot heat engine** has the maximum possible thermal efficiency of any heat engine operating between T_H and T_C:

$$\eta_{Carnot} = 1 - \frac{T_C}{T_H}$$

- A **Carnot refrigerator** has the maximum possible coefficient of performance of any refrigerator operating between T_H and T_C:

$$K_{Carnot} = \frac{T_C}{T_H - T_C}$$

The **Carnot cycle** for a gas engine consists of two isothermal processes and two adiabatic processes.

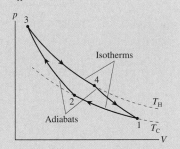

An **energy reservoir** is a part of the environment so large in comparison to the system that its temperature doesn't change as the system extracts heat energy from or exhausts heat energy to the reservoir. All heat engines and refrigerators operate between two energy reservoirs at different temperatures T_H and T_C.

The **work** W_s done *by* the system has the opposite sign to the work done *on* the system.

W_s = area under pV curve

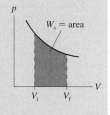

Applications

To analyze a heat engine or refrigerator:

MODEL Identify each process in the cycle.

VISUALIZE Draw the pV diagram of the cycle.

SOLVE There are several steps:

- Determine p, V, and T at the beginning and end of each process.
- Calculate ΔE_{th}, W_s, and Q for each process.
- Determine W_{in} or W_{out}, Q_H, and Q_C.
- Calculate $\eta = W_{out}/Q_H$ or $K = Q_C/W_{in}$.

ASSESS Verify $(\Delta E_{th})_{net} = 0$. Check signs.

Terms and Notation

thermodynamics	closed-cycle device	coefficient of performance, K	Carnot engine
energy reservoir	thermal efficiency, η	heat exchanger	Carnot cycle
energy-transfer diagram	waste heat	pressure ratio, r_p	
heat engine	refrigerator	perfectly reversible engine	

 For homework assigned on MasteringPhysics, go to www.masteringphysics.com
Problem difficulty is labeled as | (straightforward) to ||| (challenging).

Problems labeled ▨ integrate significant material from earlier chapters.

CONCEPTUAL QUESTIONS

1. In going from i to f in each of the three processes of **FIGURE Q19.1**, is work done *by* the system ($W < 0$, $W_s > 0$), is work done *on* the system ($W > 0$, $W_s < 0$), or is *no* net work done?

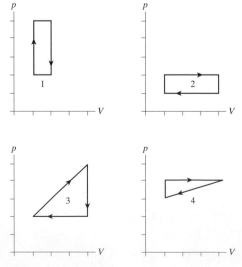

FIGURE Q19.1

2. Rank in order, from largest to smallest, the amount of work $(W_s)_1$ to $(W_s)_4$ done by the gas in each of the cycles shown in **FIGURE Q19.2**. Explain.

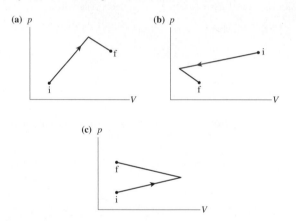

FIGURE Q19.2

3. Rank in order, from largest to smallest, the thermal efficiencies η_1 to η_4 of the four heat engines in **FIGURE Q19.3**. Explain.

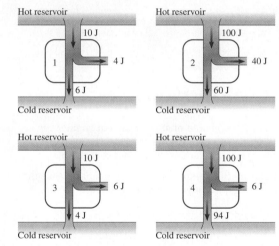

FIGURE Q19.3

4. Could you have a heat engine with $\eta > 1$? Explain.

5. **FIGURE Q19.5** shows the pV diagram of a heat engine. During which stage or stages is (a) heat added to the gas, (b) heat removed from the gas, (c) work done on the gas, and (d) work done by the gas?

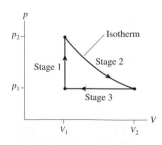

FIGURE Q19.5

6. **FIGURE Q19.6** shows the thermodynamic cycles of two heat engines. Which heat engine has the larger thermal efficiency? Or are they the same? Explain.

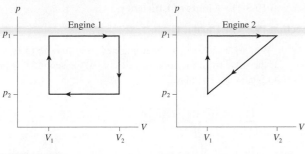

FIGURE Q19.6

7. A heat engine satisfies $W_{out} = Q_{net}$. Why is there no ΔE_{th} term in this relationship?

8. Do the energy-transfer diagrams in **FIGURE Q19.8** represent possible heat engines? If not, what is wrong?

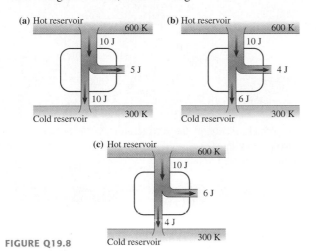

FIGURE Q19.8

9. Do the energy-transfer diagrams in **FIGURE Q19.9** represent possible refrigerators? If not, what is wrong?

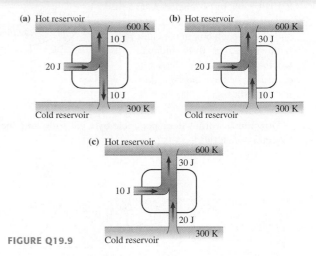

FIGURE Q19.9

10. It gets pretty hot in your apartment. In browsing the Internet, you find a company selling small "room air conditioners." You place the air conditioner on the floor, plug it in, and—the advertisement says—it will lower the room temperature up to 10°F. Should you order one? Explain.

11. The first and second laws of thermodynamics are sometimes stated as "You can't win" and "You can't even break even." Do these sayings accurately characterize the laws of thermodynamics as applied to heat engines? Why or why not?

EXERCISES AND PROBLEMS

Exercises

Section 19.1 Turning Heat into Work

Section 19.2 Heat Engines and Refrigerators

1. | A heat engine with a thermal efficiency of 40% does 100 J of work per cycle. How much heat is (a) extracted from the hot reservoir and (b) exhausted to the cold reservoir per cycle?

2. ‖ A heat engine does 20 J of work per cycle while exhausting 30 J of waste heat. What is the engine's thermal efficiency?

3. ‖ A heat engine extracts 55 kJ of heat from the hot reservoir each cycle and exhausts 40 kJ of heat. What are (a) the thermal efficiency and (b) the work done per cycle?

4. ‖ A refrigerator requires 20 J of work and exhausts 50 J of heat per cycle. What is the refrigerator's coefficient of performance?

5. | 50 J of work are done per cycle on a refrigerator with a coefficient of performance of 4.0. How much heat is (a) extracted from the cold reservoir and (b) exhausted to the hot reservoir per cycle?

6. ‖ The power output of a car engine running at 2400 rpm is 500 kW. How much (a) work is done and (b) heat is exhausted per cycle if the engine's thermal efficiency is 20%? Give your answers in kJ.

7. ‖ A 32%-efficient electric power plant produces 900 MW of electric power and discharges waste heat into 20°C ocean water. Suppose the waste heat could be used to heat homes during the winter instead of being discharged into the ocean. A typical American house requires an average 20 kW for heating. How many homes could be heated with the waste heat of this one power plant?

8. | 1.0 L of 20°C water is placed in a refrigerator. The refrigerator's motor must supply an extra 8.0 W power to chill the water to 5°C in 1.0 hr. What is the refrigerator's coefficient of performance?

Section 19.3 Ideal-Gas Heat Engines

Section 19.4 Ideal-Gas Refrigerators

9. ‖ The cycle of **FIGURE EX19.9** consists of four processes. Make a chart with rows labeled A to D and columns labeled ΔE_{th}, W_s, and Q. Fill each box in the chart with +, −, or 0 to indicate whether the quantity increases, decreases, or stays the same during that process.

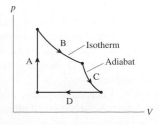

FIGURE EX19.9

10. ‖ The cycle of **FIGURE EX19.10** consists of three processes. Make a chart with rows labeled A–C and columns labeled ΔE_{th}, W_s, and Q. Fill each box in the chart with +, −, or 0 to indicate whether the quantity increases, decreases, or stays the same during that process.

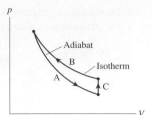

FIGURE EX19.10

11. | How much work is done per cycle by a gas following the pV trajectory of **FIGURE EX19.11**?

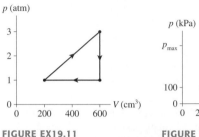

FIGURE EX19.11

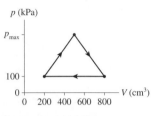

FIGURE EX19.12

12. ‖ A gas following the pV trajectory of **FIGURE EX19.12** does 60 J of work per cycle. What is p_{max}?

13. | What are (a) W_{out} and Q_C and (b) the thermal efficiency for the heat engine shown in **FIGURE EX19.13**?

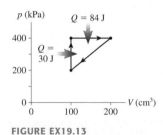

FIGURE EX19.13

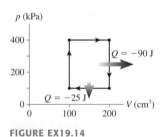

FIGURE EX19.14

14. | What are (a) W_{out} and Q_H and (b) the thermal efficiency for the heat engine shown in **FIGURE EX19.14**?

15. ‖ How much heat is exhausted to the cold reservoir by the heat engine shown in **FIGURE EX19.15**?

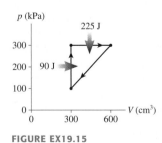

FIGURE EX19.15

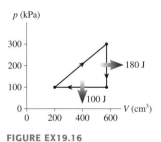

FIGURE EX19.16

16. ‖ What are (a) the thermal efficiency and (b) the heat extracted from the hot reservoir for the heat engine shown in **FIGURE EX19.16**?

17. | At what pressure ratio would a heat engine operating with a Brayton cycle have an efficiency of 60%? Assume that the gas is diatomic.

18. ‖ A heat engine uses a diatomic gas in a Brayton cycle. What is the engine's thermal efficiency if the gas volume is halved during the compression?

Section 19.5 The Limits of Efficiency

Section 19.6 The Carnot Cycle

19. | Which, if any, of the heat engines in **FIGURE EX19.19** violate (a) the first law of thermodynamics or (b) the second law of thermodynamics? Explain.

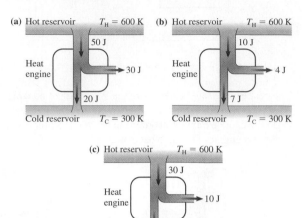

FIGURE EX19.19

20. | Which, if any, of the refrigerators in **FIGURE EX19.20** violate (a) the first law of thermodynamics or (b) the second law of thermodynamics? Explain.

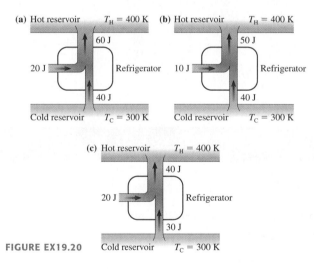

FIGURE EX19.20

21. | At what cold-reservoir temperature (in °C) would a Carnot engine with a hot-reservoir temperature of 427°C have an efficiency of 60%?

22. ‖ a. A heat engine does 200 J of work per cycle while exhausting 600 J of heat to the cold reservoir. What is the engine's thermal efficiency?

   b. A Carnot engine with a hot-reservoir temperature of 400°C has the same thermal efficiency. What is the cold-reservoir temperature in °C?

23. ‖ A heat engine does 10 J of work and exhausts 15 J of waste heat during each cycle.

   a. What is the engine's thermal efficiency?

   b. If the cold-reservoir temperature is 20°C, what is the minimum possible temperature in °C of the hot reservoir?

24. || A Carnot engine operating between energy reservoirs at temperatures 300 K and 500 K produces a power output of 1000 W. What are (a) the thermal efficiency of this engine, (b) the rate of heat input, in W, and (c) the rate of heat output, in W?

25. | A Carnot engine whose hot-reservoir temperature is 400°C has a thermal efficiency of 40%. By how many degrees should the temperature of the cold reservoir be decreased to raise the engine's efficiency to 60%?

26. || A heat engine operating between energy reservoirs at 20°C and 600°C has 30% of the maximum possible efficiency. How much energy must this engine extract from the hot reservoir to do 1000 J of work?

27. | A Carnot refrigerator operating between −20°C and +20°C extracts heat from the cold reservoir at the rate 200 J/s. What are (a) the coefficient of performance of this refrigerator, (b) the rate at which work is done on the refrigerator, and (c) the rate at which heat is exhausted to the hot side?

28. || A heat engine operating between a hot reservoir at 500°C and a cold reservoir at 0°C is 60% as efficient as a Carnot engine. If this heat engine and the Carnot engine do the same amount of work, what is the ratio $Q_H/(Q_H)_{Carnot}$?

29. || The coefficient of performance of a refrigerator is 5.0. The compressor uses 10 J of energy per cycle.
   a. How much heat energy is exhausted per cycle?
   b. If the hot-reservoir temperature is 27°C, what is the lowest possible temperature in °C of the cold reservoir?

30. || A Carnot refrigerator with a cold-reservoir temperature of −13°C has a coefficient of performance of 5.0. To increase the coefficient of performance to 10, should the hot-reservoir temperature be increased or decreased, and by how much? Explain.

Problems

31. || The engine that powers a crane burns fuel at a flame temperature of 2000°C. It is cooled by 20°C air. The crane lifts a 2000 kg steel girder 30 m upward. How much heat energy is transferred to the engine by burning fuel if the engine is 40% as efficient as a Carnot engine?

32. ||| 100 mL of water at 15°C is placed in the freezer compartment of a refrigerator with a coefficient of performance of 4.0. How much heat energy is exhausted into the room as the water is changed to ice at −15°C?

33. || Prove that the work done in an adiabatic process i → f is $W_s = (p_f V_f − p_i V_i)/(1 − \gamma)$.

34. || The hot reservoir of a heat engine is steam at 100°C while the cold reservoir is ice at 0°C. In 1.0 hr of operation, 10 kg of steam condenses and 55 kg of ice melts. What is the power output of the heat engine?

35. || Prove that the coefficient of performance of a Carnot refrigerator is $K_{Carnot} = T_C/(T_H − T_C)$.

36. || A Carnot heat engine with thermal efficiency $\frac{1}{3}$ is run backward as a Carnot refrigerator. What is the refrigerator's coefficient of performance?

37. || An ideal refrigerator utilizes a Carnot cycle operating between 0°C and 25°C. To turn 10 kg of liquid water at 0°C into 10 kg of ice at 0°C, (a) how much heat is exhausted into the room and (b) how much energy must be supplied to the refrigerator?

38. || There has long been an interest in using the vast quantities of thermal energy in the oceans to run heat engines. A heat engine needs a temperature *difference,* a hot side and a cold side. Conveniently, the ocean surface waters are warmer than the deep ocean waters. Suppose you build a floating power plant in the tropics where the surface water temperature is ≈30°C. This would be the hot reservoir of the engine. For the cold reservoir, water would be pumped up from the ocean bottom where it is always ≈5°C. What is the maximum possible efficiency of such a power plant?

39. || The ideal gas in a Carnot engine extracts 1000 J of heat energy during the isothermal expansion at 300°C. How much heat energy is exhausted during the isothermal compression at 50°C?

40. | The hot-reservoir temperature of a Carnot engine with 25% efficiency is 80°C higher than the cold-reservoir temperature. What are the reservoir temperatures, in °C?

41. || A Carnot heat engine takes 98 cycles to lift a 10 kg mass a height of 10 m. The engine exhausts 15 J of heat per cycle to a cold reservoir at 0°C. What is the temperature of the hot reservoir?

42. || The heat exhausted to the cold reservoir of a Carnot engine is two-thirds the heat extracted from the hot reservoir. What is the temperature ratio T_C/T_H?

43. || **FIGURE P19.43** shows a Carnot heat engine driving a Carnot refrigerator.
   a. Determine Q_1, Q_2, Q_3, and Q_4.
   b. Is Q_3 greater than, less than, or equal to Q_1?
   c. Do these two devices, when operated together in this way, violate the second law?

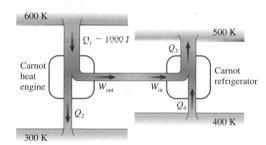

FIGURE P19.43

44. ||| A heat engine running backward is called a refrigerator if its purpose is to extract heat from a cold reservoir. The same engine running backward is called a *heat pump* if its purpose is to exhaust warm air into the hot reservoir. Heat pumps are widely used for home heating. You can think of a heat pump as a refrigerator that is cooling the already cold outdoors and, with its exhaust heat Q_H, warming the indoors. Perhaps this seems a little silly, but consider the following. Electricity can be directly used to heat a home by passing an electric current through a heating coil. This is a direct, 100% conversion of work to heat. That is, 15 kW of electric power (generated by doing work at the rate 15 kJ/s at the power plant) produces heat energy inside the home at a rate of 15 kJ/s. Suppose that the neighbor's home has a heat pump with a coefficient of performance of 5.0, a realistic value.
   a. How much electric power (in kW) does the heat pump use to deliver 15 kJ/s of heat energy to the house?
   b. An average price for electricity is about 40 MJ per dollar. A furnace or heat pump will run typically 200 hours per month during the winter. What does one month's heating cost in the home with a 15 kW electric heater and in the home of the neighbor who uses an equivalent heat pump?

45. || You and your roommates need a new refrigerator. At the appliance store, the salesman shows you the DreamFridge. According to its sticker, the DreamFridge uses a mere 100 W of power to remove 100 kJ of heat per minute from the 2°C interior. According to the fine print on the sticker, this claim is true in a 22°C kitchen. Should you buy? Explain.

46. || Three engineering students submit their solutions to a design problem in which they were asked to design an engine that operates between temperatures 300 K and 500 K. The heat input/output and work done by their designs are shown in the following table:

Student	Q_H	Q_C	W_{out}
1	250 J	140 J	110 J
2	250 J	170 J	90 J
3	250 J	160 J	90 J

Critique their designs. Are they acceptable or not? Is one better than the others? Explain.

47. ||| A typical coal-fired power plant burns 300 metric tons of coal *every hour* to generate 750 MW of electricity. 1 metric ton = 1000 kg. The density of coal is 1500 kg/m³ and its heat of combustion is 28 MJ/kg. Assume that *all* heat is transferred from the fuel to the boiler and that *all* the work done in spinning the turbine is transformed into electric energy.
    a. Suppose the coal is piled up in a 10 m × 10 m room. How tall must the pile be to operate the plant for one day?
    b. What is the power plant's thermal efficiency?

48. || A nuclear power plant generates 2000 MW of heat energy from nuclear reactions in the reactor's core. This energy is used to boil water and produce high-pressure steam at 300°C. The steam spins a turbine, which produces 700 MW of electric power, then the steam is condensed and the water is cooled to 30°C before starting the cycle again.
    a. What is the maximum possible thermal efficiency of the power plant?
    b. What is the plant's actual efficiency?
    c. Cooling water from a river flows through the condenser (the low-temperature heat exchanger) at the rate of 1.2 × 10⁸ L/hr (≈30 million gallons per hour). If the river water enters the condenser at 18°C, what is its exit temperature?

49. || The electric output of a power plant is 750 MW. Cooling water flows through the power plant at the rate 1.0 × 10⁸ L/hr. The cooling water enters the plant at 16°C and exits at 27°C. What is the power plant's thermal efficiency?

50. || a. A large nuclear power plant has a power output of 1000 MW. In other words, it generates electric energy at the rate 1000 MJ/s. How much energy does this power plant supply in one day?
    b. The oceans are vast. How much energy could be extracted from 1 km³ of water if its temperature were decreased by 1°C? For simplicity, assume fresh water.
    c. A friend of yours who is an inventor comes to you with an idea. He has done the calculations that you just did in parts a and b, and he's concluded that a few cubic kilometers of ocean water could meet most of the energy needs of the United States. This is an insignificant fraction of the U.S. coastal waters. In addition, the oceans are constantly being reheated by the sun, so energy obtained from the ocean is essentially solar energy. He has sketched out

some design plans—highly secret, of course, because they're not patented—and now he needs some investors to provide money for a prototype. A working prototype will lead to a patent. As an initial investor, you'll receive a fraction of all future royalties. Time is of the essence because a rival inventor is working on the same idea. He needs $10,000 from you right away. You could make millions if it works out. Will you invest? If so, explain why. If not, why not? Either way, your explanation should be based on scientific principles. Sketches and diagrams are a reasonable part of an explanation.

51. || An air conditioner removes 5.0 × 10⁵ J/min of heat from a house and exhausts 8.0 × 10⁵ J/min to the hot outdoors.
    a. How much power does the air conditioner's compressor require?
    b. What is the air conditioner's coefficient of performance?

52. || A heat engine using 1.0 mol of a monatomic gas follows the cycle shown in **FIGURE P19.52**. 3750 J of heat energy is transferred to the gas during process $1 \rightarrow 2$.
    a. Determine W_s, Q, and ΔE_{th} for each of the four processes in this cycle. Display your results in a table.
    b. What is the thermal efficiency of this heat engine?

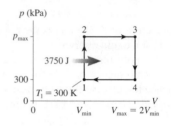

FIGURE P19.52

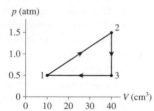

FIGURE P19.53

53. || A heat engine using a diatomic gas follows the cycle shown in **FIGURE P19.53**. Its temperature at point 1 is 20°C.
    a. Determine W_s, Q, and ΔE_{th} for each of the three processes in this cycle. Display your results in a table.
    b. What is the thermal efficiency of this heat engine?
    c. What is the power output of the engine if it runs at 500 rpm?

54. || **FIGURE P19.54** shows the cycle for a heat engine that uses a gas having $\gamma = 1.25$. The initial temperature is $T_1 = 300$ K, and this engine operates at 20 cycles per second.
    a. What is the power output of the engine?
    b. What is the engine's thermal efficiency?

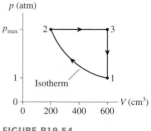

FIGURE P19.54

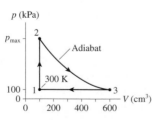

FIGURE P19.55

55. || A heat engine using a monatomic gas follows the cycle shown in **FIGURE P19.55**.
    a. Find W_s, Q, and ΔE_{th} for each process in the cycle. Display your results in a table.
    b. What is the thermal efficiency of this heat engine?

56. ‖ A heat engine uses a diatomic gas that follows the pV cycle in **FIGURE P19.56**.
   a. Determine the pressure, volume, and temperature at point 2.
   b. Determine ΔE_{th}, W_s, and Q for each of the three processes. Put your results in a table for easy reading.
   c. How much work does this engine do per cycle and what is its thermal efficiency?

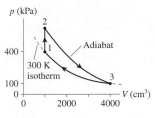

FIGURE P19.56

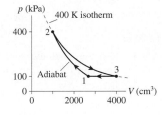

FIGURE P19.57

57. ‖ A heat engine uses a diatomic gas that follows the pV cycle in **FIGURE P19.57**.
   a. Determine the pressure, volume, and temperature at point 1.
   b. Determine ΔE_{th}, W_s, and Q for each of the three processes. Put your results in a table for easy reading.
   c. How much work does this engine do per cycle and what is its thermal efficiency?

58. ‖ A Brayton-cycle heat engine follows the cycle shown in **FIGURE P19.58**. The heat input from the burning fuel is 2.0 MJ per cycle. Determine the engine's thermal efficiency by explicitly computing the work done per cycle. Compare your answer with the efficiency that you can determine from Equation 19.21.

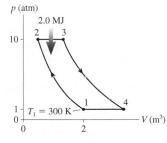

FIGURE P19.58

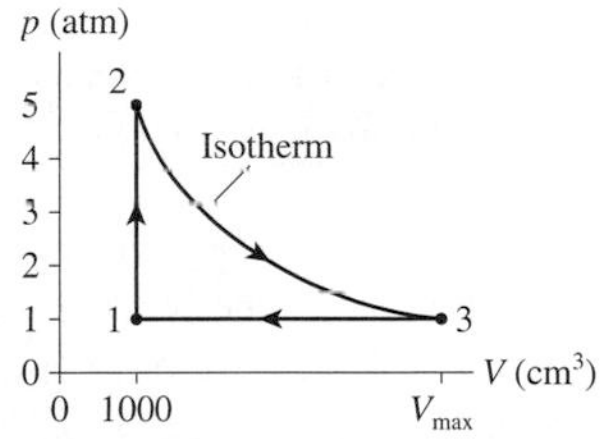

FIGURE P19.59

59. ‖ A heat engine using 120 mg of helium as the working substance follows the cycle shown in **FIGURE P19.59**.
   a. Determine the pressure, temperature, and volume of the gas at points 1, 2, and 3.
   b. What is the engine's thermal efficiency?
   c. What is the maximum possible efficiency of a heat engine that operates between T_{max} and T_{min}?

60. ‖ The heat engine shown in **FIGURE P19.60** uses 2.0 mol of a monatomic gas as the working substance.
   a. Determine T_1, T_2, and T_3.
   b. Make a table that shows ΔE_{th}, W_s, and Q for each of the three processes.
   c. What is the engine's thermal efficiency?

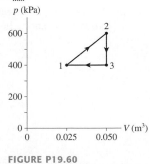

FIGURE P19.60

61. ‖‖ The heat engine shown in **FIGURE P19.61** uses 0.020 mol of a diatomic gas as the working substance.
   a. Determine T_1, T_2, and T_3.
   b. Make a table that shows ΔE_{th}, W_s, and Q for each of the three processes.
   c. What is the engine's thermal efficiency?

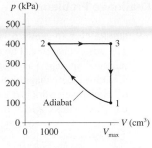

FIGURE P19.61

62. ‖‖ A heat engine using 2.0 g of helium gas is initially at STP. The gas goes through the following closed cycle:
   ■ Isothermal compression until the volume is halved.
   ■ Isobaric expansion until the volume is restored to its initial value.
   ■ Isochoric cooling until the pressure is restored to its initial value.
   How much work does this engine do per cycle and what is its thermal efficiency?

63. ‖ A heat engine with 0.20 mol of a monatomic ideal gas initially fills a 2000 cm^3 cylinder at 600 K. The gas goes through the following closed cycle:
   ■ Isothermal expansion to 4000 cm^3.
   ■ Isochoric cooling to 300 K.
   ■ Isothermal compression to 2000 cm^3.
   ■ Isochoric heating to 600 K.
   How much work does this engine do per cycle and what is its thermal efficiency?

64. ‖ **FIGURE P19.64** is the pV diagram of Example 19.2, but now the device is operated in reverse.
   a. During which processes is heat transferred into the gas?
   b. Is this Q_H, heat extracted from a hot reservoir, or Q_C, heat extracted from a cold reservoir? Explain.
   c. Determine the values of Q_H and Q_C.

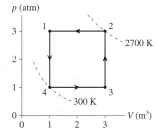

FIGURE P19.64

Hint: The calculations have been done in Example 19.2 and do not need to be repeated. Instead, you need to determine which processes now contribute to Q_H and which to Q_C.
   d. Is the area inside the curve W_{in} or W_{out}? What is its value?
   e. Show that Figure 19.19 is the energy-transfer diagram of this device.
   f. The device is now being operated in a ccw cycle. Is it a refrigerator? Explain.

In Problems 65 through 68 you are given the equation(s) used to solve a problem. For each of these, you are to
   a. Write a realistic problem for which this is the correct equation(s).
   b. Finish the solution of the problem.

65. $0.80 = 1 - (0°C + 273)/(T_H + 273)$

66. $4.0 = Q_C/W_{in}$
   $Q_H = 100$ J

67. $0.20 = 1 - Q_C/Q_H$
   $W_{out} = Q_H - Q_C = 20$ J

68. 400 kJ $= \frac{1}{2}(p_{max} - 100$ kPa$)(3.0$ m$^3 - 1.0$ m$^3)$

Challenge Problems

69. **FIGURE CP19.69** shows a heat engine going through one cycle. The gas is diatomic. The masses are such that when the pin is removed, in steps 3 and 6, the piston does not move.
    a. Draw the pV diagram for this heat engine.
    b. How much work is done per cycle?
    c. What is this engine's thermal efficiency?

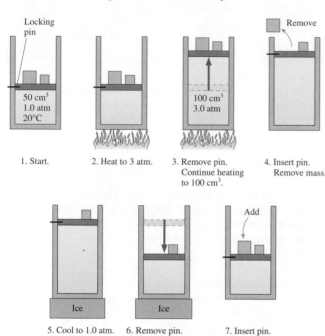

1. Start.   2. Heat to 3 atm.   3. Remove pin.
Continue heating
to 100 cm³.   4. Insert pin.
Remove mass.

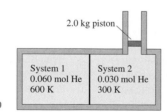

5. Cool to 1.0 atm.   6. Remove pin.
Continue cooling
to 50 cm³.   7. Insert pin.
Add mass.
Start again.

FIGURE CP19.69

70. **FIGURE CP19.70** shows two insulated compartments separated by a thin wall. The left side contains 0.060 mol of helium at an initial temperature of 600 K and the right side contains 0.030 mol of helium at an initial temperature of 300 K. The compartment on the right is attached to a vertical cylinder, above which the air pressure is 1.0 atm. A 10-cm-diameter, 2.0 kg piston can slide without friction up and down the cylinder. Neither the cylinder diameter nor the volumes of the compartments are known.
    a. What is the final temperature?
    b. How much heat is transferred from the left side to the right side?
    c. How high is the piston lifted due to this heat transfer?
    d. What fraction of the heat is converted into work?

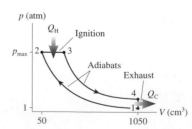

FIGURE CP19.70

71. The gasoline engine in your car can be modeled as the Otto cycle shown in **FIGURE CP19.71**. A fuel-air mixture is sprayed into the cylinder at point 1, where the piston is at its farthest distance from the spark plug. This mixture is compressed as the piston moves toward the spark plug during the adiabatic *compression stroke*. The spark plug fires at point 2, releasing heat energy that

had been stored in the gasoline. The fuel burns so quickly that the piston doesn't have time to move, so the heating is an isochoric process. The hot, high-pressure gas then pushes the piston outward during the *power stroke*. Finally, an exhaust value opens to allow the gas temperature and pressure to drop back to their initial values before starting the cycle over again.
    a. Analyze the Otto cycle and show that the work done per cycle is

$$W_{out} = \frac{nR}{1 - \gamma}(T_2 - T_1 + T_4 - T_3)$$

    b. Use the adiabatic connection between T_1 and T_2 and also between T_3 and T_4 to show that the thermal efficiency of the Otto cycle is

$$\eta = 1 - \frac{1}{r^{(\gamma - 1)}}$$

    where $r = V_{max}/V_{min}$ is the engine's *compression ratio*.
    c. Graph η versus r out to $r = 30$ for a diatomic gas.

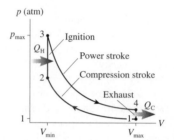

FIGURE CP19.71

72. **FIGURE CP19.72** shows the Diesel cycle. It is similar to the Otto cycle (see Problem CP19.71), but there are two important differences. First, the fuel is not admitted until the air is fully compressed at point 2. Because of the high temperature at the end of an adiabatic compression, the fuel begins to burn spontaneously. (There are no spark plugs in a diesel engine!) Second, combustion takes place more slowly, with fuel continuing to be injected. This makes the ignition stage a constant-pressure process. The cycle shown, for one cylinder of a diesel engine, has a *displacement* $V_{max} - V_{min}$ of 1000 cm³ and a compression ratio $r = V_{max}/V_{min} = 21$. These are typical values for a diesel truck. The engine operates with intake air ($\gamma = 1.40$) at 25°C and 1.0 atm pressure. The quantity of fuel injected into the cylinder has a heat of combustion of 1000 J.
    a. Find p, V, and T at each of the four corners of the cycle. Display your results in a table.
    b. What is the net work done by the cylinder during one full cycle?
    c. What is the thermal efficiency of this engine?
    d. What is the power output in kW and horsepower (1 hp = 746 W) of an eight-cylinder diesel engine running at 2400 rpm?

FIGURE CP19.72

Stop to Think 19.1: $W_d > W_a = W_b > W_c$. $W_{out} = Q_H - Q_C$.

Stop to Think 19.2: b. Energy conservation requires $Q_H = Q_C + W_{in}$. The refrigerator will exhaust more heat out the back than it removes from the front. A refrigerator with an open door will heat the room rather than cool it.

Stop to Think 19.3: c. W_{out} = area inside triangle = 1000 J. $\eta = W_{out}/Q_H = (1000 \text{ J})/(4000 \text{ J}) = 0.25$.

Stop to Think 19.4: To conserve energy, the heat Q_H exhausted to the hot reservoir needs to be $Q_H = Q_C + W_{out} = 40 \text{ J} + 10 \text{ J} = 50 \text{ J}$. The numbers shown here, with $Q_C = Q_H + W_{out}$, would be appropriate to a heat engine, not a refrigerator.

Stop to Think 19.5: b. The efficiency of this engine would be $\eta = W_{out}/Q_H = 0.6$. That exceeds the Carnot efficiency $\eta_{Carnot} = 1 - T_C/T_H = 0.5$, so it is not possible.

Thermodynamics

Part IV had two important goals: first, to learn how energy is transformed; second, to establish a micro/macro connection in which we can understand the macroscopic properties of solids, liquids, and gases in terms of the microscopic motions of atoms and molecules. We have been quite successful. You have learned that:

- Temperature is a measure of the thermal energy of the molecules in a system, and the average energy per molecule is simply $\frac{1}{2}k_BT$ per degree of freedom.
- The pressure of a gas is due to collisions of the molecules with the walls of the container.
- Heat is the energy transferred between two systems that have different temperatures. The mechanism of heat transfer is molecular collisions at the boundary between the two systems.
- Work, heat, and thermal energy can be transformed into each other in accord with the first law of thermodynamics, $\Delta E_{th} = W + Q$. This is a statement that energy is conserved.

- Practical devices for turning heat into work, called heat engines, are limited in their efficiency by the second law of thermodynamics.

The knowledge structure of thermodynamics below summarizes the basic laws, diagramming our energy model and presenting our model of a heat engine in pictorial form. Thermodynamics, more than most topics in physics, can seem very "equation oriented." It's undeniable that there are more equations than we used in earlier parts of this text and more things to remember. But focusing on the equations is seeing only the trees, not the forest. A better strategy is to focus on the ideas embedded in the knowledge structure. You can find the necessary equations if you know how the ideas are connected, but memorizing all the equations won't help if you don't know which are relevant to different situations.

KNOWLEDGE STRUCTURE IV **Thermodynamics**

ESSENTIAL CONCEPTS	Work, heat, and thermal energy.
BASIC GOALS	How is energy converted from one form to another?
	How are macroscopic properties related to microscopic behavior?

GENERAL PRINCIPLES	**First law of thermodynamics**	Energy is conserved, $\Delta E_{th} = W + Q$.
	Second law of thermodynamics	Heat is not spontaneously transferred from a colder object to a hotter object.

GAS LAWS AND PROCESSES   Ideal-gas law $pV = nRT = Nk_BT$

- Isochoric process $V = $ constant and $W = 0$
- Isothermal process $T = $ constant and $\Delta E_{th} = 0$

- Isobaric process $p = $ constant
- Adiabatic process $Q = 0$

Energy Transformation

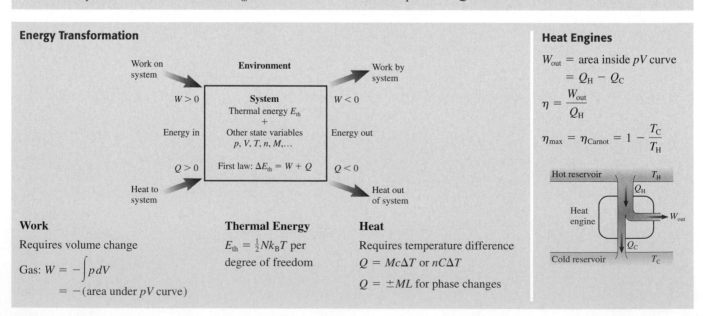

Heat Engines

$W_{out} = $ area inside pV curve

$\quad\quad = Q_H - Q_C$

$\eta = \dfrac{W_{out}}{Q_H}$

$\eta_{max} = \eta_{Carnot} = 1 - \dfrac{T_C}{T_H}$

Work

Requires volume change

Gas: $W = -\displaystyle\int p\,dV$

$\quad = -$ (area under pV curve)

Thermal Energy

$E_{th} = \frac{1}{2}Nk_BT$ per degree of freedom

Heat

Requires temperature difference

$Q = Mc\Delta T$ or $nC\Delta T$

$Q = \pm ML$ for phase changes

Order Out of Chaos

The second law predicts that systems will run down, that order will evolve toward disorder and randomness, and that complexity will give way to simplicity. But just look around you!

- Plants grow from simple seeds to complex entities.
- Single-cell fertilized eggs grow into complex adult organisms.
- Electric current passing through a "soup" of simple random molecules produces such complex chemicals as amino acids.
- Over the last billion or so years life has evolved from simple unicellular organisms to very complex forms.
- Knowledge and information seem to grow every year, not to fade away.

Everywhere we look, it seems, the second law is being violated. How can this be?

There is an important qualification in the second law of thermodynamics: It applies only to *isolated* systems, systems that do not exchange energy with their environment. The situation is entirely different if energy is transferred into or out of the system, and we cannot predict what will happen to the entropy of a nonisolated system. The popular-science literature is full of arguments and predictions that make incorrect use of the second law by trying to apply it to systems that are not isolated.

Systems that become *more* ordered as time passes, and in which the entropy decreases, are called *self-organizing systems.* All the examples listed above are self-organizing systems. One of the major characteristics of self-organizing systems is a substantial flow of energy *through* the system. For example, plants and animals take in energy from the sun or chemical energy from food, make use of that energy, and then give waste heat back to the environment via evaporation, decay, and other means. It is this energy flow that allows the systems to maintain, or even increase, a high degree of order and a very low entropy.

But—and this is the important point—the entropy of the *entire* system, including the earth and the sun, undergoes a significant *increase* so as to let selected subsystems decrease their entropy and become more ordered. The second law is not violated at all, but you must apply the second law to the combined systems that are interacting and not just to a single subsystem.

The snowflake in the photo is a beautiful example. As water freezes, the random motion of water molecules is transformed into a highly ordered crystal. The entropy of

A snowflake is a highly ordered arrangement of water molecules. The creation of a snowflake decreases the entropy of the water, but the second law of thermodynamics is not violated because the water molecules are not an isolated system.

the water molecules certainly decreases, but water doesn't freeze as an isolated system. For it to freeze, heat energy must be transferred from the water to the surrounding air. The entropy of the air increases by *more* than the entropy of the water decreases. Thus the *total* entropy of the water + air system increases when a snowflake is formed, just as the second law predicts.

Self-organization is closely related to nonlinear mechanics, chaos, and the geometry of fractals. It has important applications in fields ranging from ecology to computer science to aeronautical engineering. For example, the airflow across a wing gives rise to large-scale turbulence—eddies and whirlpools—in the wake behind an airplane. Their formation affects the aerodynamics of the plane and can also create hazards for following aircraft. Whirlpools are ordered, large-scale macroscopic structures with low entropy, but they are produced from disordered, random collisions of the air molecules.

Self-organizing systems are a very active field of research in both science and engineering. The 1977 Nobel Prize in chemistry was awarded to the Belgian scientist Ilya Prigogine for his studies of *nonequilibrium thermodynamics,* the basic science underlying self-organizing systems. Prigogine and others have shown how energy flow through a system can, when the conditions are right, "bring order out of chaos."

Mathematics Review

Algebra

Using exponents:
$$a^{-x} = \frac{1}{a^x} \qquad a^x a^y = a^{(x+y)} \qquad \frac{a^x}{a^y} = a^{(x-y)} \qquad (a^x)^y = a^{xy}$$

$$a^0 = 1 \qquad a^1 = a \qquad a^{1/n} = \sqrt[n]{a}$$

Fractions:
$$\left(\frac{a}{b}\right)\left(\frac{c}{d}\right) = \frac{ac}{bd} \qquad \frac{a/b}{c/d} = \frac{ad}{bc} \qquad \frac{1}{1/a} = a$$

Logarithms:
If $a = e^x$, then $\ln(a) = x$ $\qquad \ln(e^x) = x \qquad e^{\ln(x)} = x$

$$\ln(ab) = \ln(a) + \ln(b) \qquad \ln\left(\frac{a}{b}\right) = \ln(a) - \ln(b) \qquad \ln(a^n) = n\ln(a)$$

The expression $\ln(a + b)$ cannot be simplified.

Linear equations: The graph of the equation $y = ax + b$ is a straight line. a is the slope of the graph. b is the y-intercept.

Proportionality: To say that y is proportional to x, written $y \propto x$, means that $y = ax$, where a is a constant. Proportionality is a special case of linearity. A graph of a proportional relationship is a straight line that passes through the origin. If $y \propto x$, then

$$\frac{y_1}{y_2} = \frac{x_1}{x_2}$$

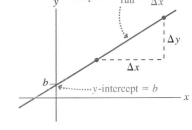

Quadratic equation: The quadratic equation $ax^2 + bx + c = 0$ has the two solutions $x = \dfrac{-b \pm \sqrt{b^2 - 4ac}}{2a}$.

Geometry and Trigonometry

Area and volume:

Rectangle
$A = ab$
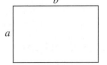

Rectangular box
$V = abc$
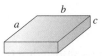

Triangle
$A = \frac{1}{2}ab$

Right circular cylinder
$V = \pi r^2 l$

Circle
$C = 2\pi r$
$A = \pi r^2$

Sphere
$A = 4\pi r^2$
$V = \frac{4}{3}\pi r^3$

Arc length and angle: The angle θ in radians is defined as $\theta = s/r$.

The arc length that spans angle θ is $s = r\theta$.

2π rad $= 360°$

Right triangle: Pythagorean theorem $c = \sqrt{a^2 + b^2}$ or $a^2 + b^2 = c^2$

$$\sin\theta = \frac{b}{c} = \frac{\text{far side}}{\text{hypotenuse}} \qquad \theta = \sin^{-1}\left(\frac{b}{c}\right)$$

$$\cos\theta = \frac{a}{c} = \frac{\text{adjacent side}}{\text{hypotenuse}} \qquad \theta = \cos^{-1}\left(\frac{a}{c}\right)$$

$$\tan\theta = \frac{b}{a} = \frac{\text{far side}}{\text{adjacent side}} \qquad \theta = \tan^{-1}\left(\frac{b}{a}\right)$$

General triangle: $\alpha + \beta + \gamma = 180° = \pi$ rad

Law of cosines $c^2 = a^2 + b^2 - 2ab\cos\gamma$

Identities:

$\tan\alpha = \dfrac{\sin\alpha}{\cos\alpha}$ $\qquad\qquad$ $\sin^2\alpha + \cos^2\alpha = 1$

$\sin(-\alpha) = -\sin\alpha$ $\qquad\qquad$ $\cos(-\alpha) = \cos\alpha$

$\sin(\alpha \pm \beta) = \sin\alpha\cos\beta \pm \cos\alpha\sin\beta$ $\qquad$ $\cos(\alpha \pm \beta) = \cos\alpha\cos\beta \mp \sin\alpha\sin\beta$

$\sin(2\alpha) = 2\sin\alpha\cos\alpha$ $\qquad\qquad$ $\cos(2\alpha) = \cos^2\alpha - \sin^2\alpha$

$\sin(\alpha \pm \pi/2) = \pm\cos\alpha$ $\qquad\qquad$ $\cos(\alpha \pm \pi/2) = \mp\sin\alpha$

$\sin(\alpha \pm \pi) = -\sin\alpha$ $\qquad\qquad$ $\cos(\alpha \pm \pi) = -\cos\alpha$

Expansions and Approximations

Binomial expansion: $(1 + x)^n = 1 + nx + \dfrac{n(n-1)}{2}x^2 + \cdots$

Binomial approximation: $(1 + x)^n \approx 1 + nx$ if $x \ll 1$

Trigonometric expansions: $\sin\alpha = \alpha - \dfrac{\alpha^3}{3!} + \dfrac{\alpha^5}{5!} - \dfrac{\alpha^7}{7!} + \cdots$ for α in rad

$\cos\alpha = 1 - \dfrac{\alpha^2}{2!} + \dfrac{\alpha^4}{4!} - \dfrac{\alpha^6}{6!} + \cdots$ for α in rad

Small-angle approximation: If $\alpha \ll 1$ rad, then $\sin\alpha \approx \tan\alpha \approx \alpha$ and $\cos\alpha \approx 1$.

The small-angle approximation is excellent for $\alpha < 5°$ (≈ 0.1 rad) and generally acceptable up to $\alpha \approx 10°$.

Calculus

The letters a and n represent constants in the following derivatives and integrals.

Derivatives

$$\frac{d}{dx}(a) = 0$$

$$\frac{d}{dx}(ax) = a$$

$$\frac{d}{dx}\left(\frac{a}{x}\right) = -\frac{a}{x^2}$$

$$\frac{d}{dx}(ax^n) = anx^{n-1}$$

$$\frac{d}{dx}(\ln(ax)) = \frac{1}{x}$$

$$\frac{d}{dx}(e^{ax}) = ae^{ax}$$

$$\frac{d}{dx}(\sin(ax)) = a\cos(ax)$$

$$\frac{d}{dx}(\cos(ax)) = -a\sin(ax)$$

Integrals

$$\int x\,dx = \frac{1}{2}x^2$$

$$\int x^2\,dx = \frac{1}{3}x^3$$

$$\int \frac{1}{x^2}\,dx = -\frac{1}{x}$$

$$\int x^n\,dx = \frac{x^{n+1}}{n+1} \qquad n \neq -1$$

$$\int \frac{dx}{x} = \ln x$$

$$\int \frac{dx}{a+x} = \ln(a+x)$$

$$\int \frac{x\,dx}{a+x} = x - a\ln(a+x)$$

$$\int \frac{dx}{\sqrt{x^2 \pm a^2}} = \ln\left(x + \sqrt{x^2 \pm a^2}\right)$$

$$\int \frac{x\,dx}{\sqrt{x^2 \pm a^2}} = \sqrt{x^2 \pm a^2}$$

$$\int \frac{dx}{x^2 + a^2} = \frac{1}{a}\tan^{-1}\left(\frac{x}{a}\right)$$

$$\int \frac{dx}{(x^2 + a^2)^2} = \frac{1}{2a^3}\tan^{-1}\left(\frac{x}{a}\right) + \frac{x}{2a^2(x^2 + a^2)}$$

$$\int \frac{dx}{(x^2 \pm a^2)^{3/2}} = \frac{\pm x}{a^2\sqrt{x^2 \pm a^2}}$$

$$\int \frac{x\,dx}{(x^2 \pm a^2)^{3/2}} = -\frac{1}{\sqrt{x^2 \pm a^2}}$$

$$\int e^{ax}\,dx = \frac{1}{a}e^{ax}$$

$$\int xe^{ax}\,dx = \frac{1}{a^2}e^{ax}(ax - 1)$$

$$\int \sin(ax)\,dx = -\frac{1}{a}\cos(ax)$$

$$\int \cos(ax)\,dx = \frac{1}{a}\sin(ax)$$

$$\int \sin^2(ax)\,dx = \frac{x}{2} - \frac{\sin(2ax)}{4a}$$

$$\int \cos^2(ax)\,dx = \frac{x}{2} + \frac{\sin(2ax)}{4a}$$

$$\int_0^\infty x^n e^{-ax}\,dx = \frac{n!}{a^{n+1}}$$

$$\int_0^\infty e^{-ax^2}\,dx = \frac{1}{2}\sqrt{\frac{\pi}{a}}$$

Periodic Table of Elements

Period																	

Atomic number — 27
Symbol — Co
Atomic mass — 58.9

Transition elements

1																	2
H 1.0																	He 4.0

| 3 Li 6.9 | 4 Be 9.0 | | | | | | | | | | | 5 B 10.8 | 6 C 12.0 | 7 N 14.0 | 8 O 16.0 | 9 F 19.0 | 10 Ne 20.2 |

| 11 Na 23.0 | 12 Mg 24.3 | | | | | | | | | | | 13 Al 27.0 | 14 Si 28.1 | 15 P 31.0 | 16 S 32.1 | 17 Cl 35.5 | 18 Ar 39.9 |

| 19 K 39.1 | 20 Ca 40.1 | 21 Sc 45.0 | 22 Ti 47.9 | 23 V 50.9 | 24 Cr 52.0 | 25 Mn 54.9 | 26 Fe 55.8 | 27 Co 58.9 | 28 Ni 58.7 | 29 Cu 63.5 | 30 Zn 65.4 | 31 Ga 69.7 | 32 Ge 72.6 | 33 As 74.9 | 34 Se 79.0 | 35 Br 79.9 | 36 Kr 83.8 |

| 37 Rb 85.5 | 38 Sr 87.6 | 39 Y 88.9 | 40 Zr 91.2 | 41 Nb 92.9 | 42 Mo 95.9 | 43 Tc 96.9 | 44 Ru 101.1 | 45 Rh 102.9 | 46 Pd 106.4 | 47 Ag 107.9 | 48 Cd 112.4 | 49 In 114.8 | 50 Sn 118.7 | 51 Sb 121.8 | 52 Te 127.6 | 53 I 126.9 | 54 Xe 131.3 |

| 55 Cs 132.9 | 56 Ba 137.3 | 57 La 138.9 | 72 Hf 178.5 | 73 Ta 180.9 | 74 W 183.9 | 75 Re 186.2 | 76 Os 190.2 | 77 Ir 192.2 | 78 Pt 195.1 | 79 Au 197.0 | 80 Hg 200.6 | 81 Tl 204.4 | 82 Pb 207.2 | 83 Bi 209.0 | 84 Po 209.0 | 85 At 210.0 | 86 Rn 222.0 |

| 87 Fr 223.0 | 88 Ra 226.0 | 89 Ac 227.0 | 104 Rf 261 | 105 Db 262 | 106 Sg 263 | 107 Bh 264 | 108 Hs 269 | 109 Mt 268 | 110 Ds 271 | 111 Rg 272 | 112 285 | | | | | | |

Lanthanides 6

Actinides 7

Inner transition elements

58 Ce 140.1	59 Pr 140.9	60 Nd 144.2	61 Pm 144.9	62 Sm 150.4	63 Eu 152.0	64 Gd 157.3	65 Tb 158.9	66 Dy 162.5	67 Ho 164.9	68 Er 167.3	69 Tm 168.9	70 Yb 173.0	71 Lu 175.0
90 Th 232.0	91 Pa 231.0	92 U 238.0	93 Np 237.0	94 Pu 239.1	95 Am 241.1	96 Cm 244.1	97 Bk 249.1	98 Cf 252.1	99 Es 257.1	100 Fm 257.1	101 Md 258.1	102 No 259.1	103 Lr 262.1

Answers

Answers to Odd-Numbered Exercises and Problems

Chapter 16

1. $22.6 \, m^3$
3. $4.2 \, cm$
5. 4.8×10^{23} atoms
7. a. 6.02×10^{28} atoms/m^3 b. 3.28×10^{28} atoms/m^3
9. $4.06 \, cm$
11. $-127°F = -88°C = 185 \, K$, $136°F = 58°C = 331 \, K$
13. a. $171°Z$ b. $944°Z$
15. $19 \, atm$
17. a. $0.050 \, m^3$ b. $1.3 \, atm$
19. a. $55 \, mol$ b. $1.2 \, m^3$
21. a. 5.4×10^{23} atoms b. $3.6 \times 10^{-3} \, kg$ c. 2.3×10^{26} atoms/m^3
    d. $1.5 \, kg/m^3$
23. a. $V_2 = V_1$ b. $T_2 = T_1/3$
25. a. $0.73 \, atm$ b. $0.52 \, atm$
27. a. $9500 \, kPa$
29. a. $48 \, atm$
31. a. Isothermal b. $641°C$ c. $300 \, cm^3$
33. $0.228 \, nm$
35. 3.3×10^{26} protons
37. 1.1×10^{15} particles/m^3
39. a. 1.3×10^{-13} b. 1.2×10^{11} molecules
41. $1.8 \, g$
43. $174°C$
45. $93 \, cm^3$
47. $35 \, psi$
49. $174.3°C$
51. $24.0 \, cm$
53. No
55.

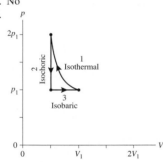

57. a. $880 \, kPa$ b. $T_2 = 323°C$, $T_3 = -49°C$, $T_4 = 398°C$
59. a. $T_1 = 366 \, K$, $T_2 = 366 \, K$ b. Isothermal c. $825°C$
61. $2364°C$
63. a. $4.0 \, atm$, $-73°C$
    b.

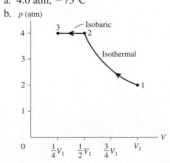

65. b.

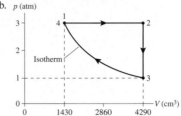

    c. $6 \, atm$
67. b.

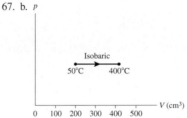

    c. $417 \, cm^3$
69. a. $23.5 \, cm$ b. $7.8 \, cm$
71. $2.4 \, m$
73. a. $4.0 \times 10^5 \, Pa$ b. Irreversible

Chapter 17

1. $40 \, J$
3. $200 \, cm^3$
5.

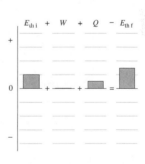

7.

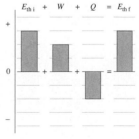

A-5

9. 700 J from system
11. 12,000 J
13. 6860 J
15. 6.8×10^4 J
17. 28°C
19. 73.5°C
21. Iron
23. a. 91 J   b. 140°C
25. a. 1.32   b. 1.25
27. a. 1.1 atm   b. 48°C
29. Iron
31. 26 W
33. 8.7 h
35. 990 cm^3
37. −56°C
39. Aluminum
41. 87 min
43. -3.4×10^5 J
45. 5500 J
47. 2.8 atm
49. 1660 J
51. a. 253°C   b. 33 cm
53. a. 110 kPa   b. 24 cm
55. a. $V_2 = 4300$ cm^3 $T_2 = 606$°C   b. 3000 J   c. 1.0 atm   d. 2200 J
   e.

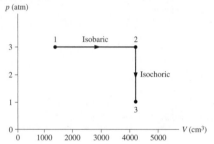

57. For A: −1000 J; for B: 1400 J
59. a. −410 J   b. 570 J   c. 0 J
61. a. $T_{Af} = 300$ K, $T_{Bf} = 220$ K, $V_{Af} = 2.5 \times 10^{-3}$ m^3,
   $V_{Bf} = 1.8 \times 10^{-3}$ m^3
   b.

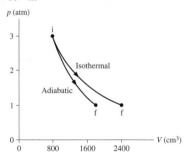

63. a. −50.7 J   b. −15 J   c. 36 J
65. $T_f = 830$°C, $V_f = 24$ cm^3
67. a. 39.3   b. 171

69. a. 0.50 atm   b. −1070 J   c. 1070 J   d. 0 J
   e.

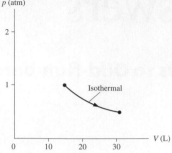

71. a. 5500 K   b. 0 J   c. 5.4×10^4 J   d. 20
   e.

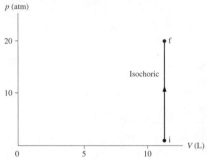

73. 110°C
75. −18°C
77. b. 217°C
79. a.

Point	p (atm)	T (°C)	V (cm^3)
1	3.0	946	1000
2	1.0	946	3000
3	0.48	310	3000

   b. $W_{1\rightarrow2} = -334$ J, $W_{2\rightarrow3} = 0$ J, $W_{3\rightarrow1} = 239$ J
   c. $Q_{1\rightarrow2} = 334$ J, $Q_{2\rightarrow3} = -239$ J, $Q_{3\rightarrow1} = 0$ J
81. 15 atm
83. a. 606°C   b. 150 J

Chapter 18

1. 2.69×10^{25} m^{-3}
3. 0.023 Pa
5. a. 300 nm   b. 600 nm
7. 12.5 cm
9. a. $(0\hat{i} + 0\hat{j})$ m/s   b. 59 m/s   c. 62 m/s
11. a. 9.16 Pa   b. 332 K
13. Neon
15. 2.5 mK
17. −246°C
19. 800 m/s
21. 7.22×10^{12} K
23. a. 3400 J   b. 3400 J   c. 3400 J
25. a. 4.1×10^{-16} J   b. 7.0×10^5 m/s
27. 490 J
29. 3.6×10^7 J
31. a. 0.080°C   b. 0.048°C   c. 0.040°C
33. a. 62 J   b. 100 J   c. 100 J   d. 150 J
35. a. B   b. $E_{Af} = 5200$ J, $E_{Bf} = 7800$ J
37. 84.8

39. a. Helium   b. 1370 m/s   c. 1.86 μm
41. 9.6×10^{-5} m/s
43. a. $\lambda_{electron} = \dfrac{1}{\sqrt{2}\pi(N/V)r^2}$

    b. 1.82×10^{-6} Pa $= 1.80 \times 10^{-11}$ atm
45. a. 1.3×10^{25} m^{-3} b. 450 m/s   c. 260 m/s   d. 1.3×10^{22} s^{-1}
    e. 57 kPa   f. 57 kPa
47. 29 J/mol K
49. a. $(E_{He})_i = 1900$ J, $(E_O)_i = 3100$ J   b. $(E_{He})_f = 2700$ J,
    $(E_O)_f = 2300$ J   c. 850 J from oxygen to helium   d. 436 K
51. 7
55. a. R b. $2R$
57. a. 4   b. 1   c. 16
59. a. 141,000 T   b. 10,100 K   c. For N$_2$ $\epsilon_{avg}/K_{esc} = 0.2\%$;
    For H$_2$ $\epsilon_{avg}/K_{esc} = 3\%$;
61. a. 2.0×10^6 J   b. 4.8×10^{-6} c. 0.0013°C
63. b. $9p_iV_i$
65. c. 436 K; 850 J is transferred from oxygen to helium

Chapter 19

1. a. 250 J   b. 150 J
3. a. 0.27   b. 14 kJ
5. a. 200 J   b. 250 J
7. 96,000
9.

	ΔE_{th}	W_s	Q
A	+	0	+
B	0	+	+
C	−	+	0
D	−	−	−

11. 20.5 J
13. a. $W_{out} = 10$ J, $Q_C = 110$ J   b. 0.083
15. 283 J
17. 25
19. a. (b)   b. (a)
21. 7°C
23. a. 0.40   b. 215°C
25. 135°C
27. a. 6.33   b. 32 W   c.   232 W
29. a. 60 J   b. −23°C
31. 1.7×10^6 J
34. 1200 W
37. a. 3.63×10^6 J   b. 3.0×10^5 J
39. 560 J
41. 1820°C
43. a. $Q_1 = 1000$ J $Q_2 = 500$ J $Q_3 = 2500$ J $Q_4 = 2000$ J
    b. $Q_3 > Q_1$ c. No
45. No
47. a. 48 m   b. 0.32

49. 0.37
51. a. 5.0 kW   b. 1.7
53. a.

	W_s (J)	Q (J)	ΔE_{th}
1 → 2	3.04	16.97	13.93
2 → 3	0	−10.13	−10.13
3 → 1	−1.52	−5.32	−3.80
Net	1.52	1.52	0

    b. 0.090   c. 13 W
55. a.

	W_s (J)	Q (J)	ΔE_{th} (J)
1 → 2	0	282.2	282.2
2 → 3	207.2	0	−207.2
3 → 1	−50.0	−125.0	−75.0
Net	157.2	157.2	0

    b. 0.52
57. a. $V_1 = 4000$ cm^3 $p_1 = 5.7$ kPa $T_1 = 230$ K
    b.

	ΔE_{th} (J)	W_s (J)	Q (J)
1 → 2	425.7	−425.7	0
2 → 3	0	554.5	554.5
3 → 1	−425.7	0	−425.7
Net	0	128.8	128.8

    c. 0.23
59. a.

	p (atm)	T (K)	V (cm^3)
1	1.0	406	1000
2	5.0	2030	1000
3	1.0	2030	5000

    b. 0.29   c. 0.80
61. a. $T_1 = 1620$ K $T_2 = 2407$ K $T_3 = 6479$ K
    b.

	ΔE_{th} (J)	W_s (J)	Q (J)
1 → 2	327	−327	0
2 → 3	1692	677	2369
3 → 1	−2019	0	−2019
Net	0	350	350

    c. 0.15
63. $W_{net} = 350$ J $\eta = 0.24$
65. b. $T_H = 1092$°C
67. b. $Q_C = 80$ J
69. b. 10.13′ J   c. 0.13

Credits

INTRODUCTION

Courtesy of International Business Machine Corporation. Unauthorized use not permitted.

TITLE PAGE

Composite illustration by Yvo Riezebos Design, photo of spring by Bill Frymire/Masterfile.

PART IV OPENER

Page **478:** Pictor International/Alamy.

CHAPTER 16

Page **480:** Rolf Hicker/Arco Digital Images. Page **481:** Esbin-Anderson/The Image Works. Page **484:** Richard Megna/Fundamental Photos. Page **485** T: David Young-Wolff/PhotoEdit. Page **485** B: David Young-Wolff/PhotoEdit. Page **489:** Tim McGuire/Corbis.

CHAPTER 17

Page **506:** Dr. Arthur Tucker/Photo Researchers. Page **509:** Robert & Linda Mostyn/Eye Ubiquitous/Corbis. Page **513:** Kevin Fleming/Corbis. Page **521:** Roger Ressmeyer/Corbis. Page **530** L: Cn Boon Alamy. Page **530** ML: Andrew Davidhazy. Page **530** MR: Pascal Goetgheluck/Photo Researchers. Page **530** R: Corbis. Page **531:** sciencephotos/Alamy Page **532:** Johns Hopkins University Applied Physics Laboratory.

CHAPTER 18

Page **541:** Gerolf Kalt/Corbis. Page **559:** Photodisc/Getty Images.

CHAPTER 19

Page **566:** Chris Ladd/Getty Images. **567:** Spencer Grant/PhotoEdit. Page **570:** Peter Bowater/Alamy. Page **573:** Peter Bowater/Alamy. Page **579:** Malcolm Fife/Getty Images. Page **580:** Paul Silverman/Fundamental Photos.

PART IV SUMMARY

Page **599:** Kristian Hilsen/Getty Images. Fundamental Photographs.

Index

For users of the five-volume edition: pages 1–477 are in Volume 1; pages 478–599 are in Volume 2; pages 600–785 for Volume 3; pages 786–1183 for Volume 4; and pages 1140–1365 are in Volume 5. Pages 1184–1365 are not in the Standard Edition.